S0-ABM-668

Heathkit

Educational
Systems

SEMICONDUCTOR DEVICES

TEXTBOOK

Revised by:

Philip Wheeler

Model EB-6103A
HEATH COMPANY
BENTON HARBOR, MICHIGAN 49022
595-3408-04

Copyright © **1978**
Revised 1985
Eleventh Printing——1988
Heath Company
Not affiliated with D.C. Heath
All Rights Reserved
Printed in the United States of America

Copyright (C) 1978 and revised 1985

Heath Company
Not affiliated with D.C. Heath Inc.
All Rights Reserved
Printed in the United States of America

II SEMICONDUCTOR DEVICES

Contents

Introduction

Semiconductor devices have revolutionized both the electronics industry and the modern world. Semiconductor, or solid-state, devices have made it possible to design circuits and equipment that were not even conceived until recently. In fact, many of the electronic devices that we now take for granted, such as hand-held calculators, were impossible to design or manufacture just a few years ago. Some devices, such as digital wrist watches, could not be made at all if it were not for semiconductors. The low cost of these devices has made them affordable to almost everyone, resulting in a high demand that encourages further innovation. One example of a major recent innovation is the microcomputer.

This course describes the most common semiconductor devices and their theory of operation. Careful study of this material will lead you to a good understanding of the types and uses of solid-state components. It is to your advantage to answer all Progress Check questions and perform all experiments supplied with the course. You can further test your understanding of semiconductor devices by taking the Unit Examination located in the Student Workbook. Most importantly, after taking the examination, review any material that was tested by a question that you answered incorrectly.

Course Objectives

When you have completed this course on semiconductor devices you will be able to:

1. Describe the electrical characteristics of materials which are classified as semiconductors.

2. Name the primary advantages that semiconductor devices have over vacuum tubes.

3. Explain how the most important semiconductor devices operate and their particular biasing requirements.

4. Basically describe how the most important semiconductor devices are constructed.

5. Handle semiconductor components properly without exceeding their maximum ratings or damaging them with improper handling procedures.

6. Recognize the most commonly used semiconductor packages.

7. Recognize the schematic symbols that are used to represent a wide variety of semiconductor devices.

8. Test various semiconductor devices to determine if they are functioning properly.

9. Use a variety of common semiconductors in practical circuits.

10. Analyze semiconductor circuits and verify proper operation using test equipment.

11. Make comparisons between solid state devices and decide which is best suited for a particular application.

12. Identify an unknown transistor as an NPN or PNP and determine which is the base lead.

13. Explain the terms linear (analog) and digital components.

14. State the advantages of using discrete components, MOS devices, and CMOS devices.

15. From a schematic determine which devices are light sensing and which devices are light producing.

Course Outline

Unit 1
Semiconductor Fundamentals

Contents

Introduction

In this unit on semiconductor fundamentals, you will learn how and why semiconductors are used in electronics. You will examine various types of semiconductor materials and study their electrical characteristics. The information presented in this unit is extremely important because it forms the basic foundation upon which you will eventually build your knowledge of semiconductor devices. Examine the Unit Objectives closely so that you will know exactly what you should achieve in this unit.

You will notice that the subject of current flow is not as simple and straight forward as it was in the AC and DC courses. This is because in semiconductor material current flow is thought of as charge carriers, and the charge carriers can be either electrons or holes. In many cases both holes and electrons are considered to move within the same component. Thus, the terms majority and minority current carriers are used to describe circuit operation.

In P-type semiconductor material, holes are the majority carriers, and in N-type material, electrons are the majority carriers. The important concept is that there is a cause and effect relationship. Since electrons and holes have an opposite charge potential (holes are + and electrons are −) and they are both current carriers. Therefore, they are both in motion, but that they move in opposite directions. This will be fully explained in this and other units. Let's start by carefully reading the unit objectives.

Unit Objectives

When you have completed this unit, you should be able to:

1. Identify three common semiconductor devices.

2. Describe some of the uses of semiconductor devices in electronic equipment.

3. List at least five advantages semiconductor devices have over other components having similar capabilities.

4. Identify the two most commonly used semiconductor materials and identify their majority and minority current carriers.

5. Describe the crystal lattice structure of semiconductors.

6. Define the electrical characteristics of semiconductors.

7. Describe the difference between intrinsic and doped semiconductors.

8. Define the term "hole" as applied to semiconductors.

9. Recognize the difference between majority carriers and minority carriers in a doped semiconductor.

10. Explain the terms trivalent and pentavalent and explain why they are used.

The Importance Of Semiconductors

Semiconductors are the basic building materials which are used to construct some very important electronic components. These semiconductor components are in turn used to construct electronic circuits and equipment. The three most commonly used semiconductor devices are diodes, transistors, and integrated circuits. However, other special components are also available. Tell me what I say

The primary function of semiconductor devices in electronic equipment is to control currents or voltages in such a way as to produce a desired end result. For example, diodes can be used as rectifiers to produce pulsating DC from AC. A transistor can be used as a variable resistance to vary the current in a heating element. Or an integrated circuit can be used to amplify and demodulate a radio signal. All of these components are made of special materials known as semiconductors.

Semiconductor devices are extremely small, lightweight components which consume only a small amount of power and are highly efficient and reliable. The vacuum tubes that were once widely used in practically all types of electronic equipment have been almost completely replaced by the newer and better semiconductor devices. Let's consider some of the specific reasons for this significant transition from the use of vacuum tubes to semiconductor components in electronic equipment.

Advantages- Not

Components which are made of semiconductor materials are often referred to as solid-state components because they are made from solid materials. Because of this solid-state construction, these components are more rugged than vacuum tubes which are made of a combination of glass, metal, and ceramic materials. Because of this ruggedness, semiconductor devices are able to operate under extremely hazardous environmental conditions. This ruggedness is responsible for the reliability of solid-state devices.

The solid-state construction also eliminates the need for filaments or heaters as found in all vacuum tubes. This means that additional power is not required to operate the filaments and component operation is cooler and more efficient. By eliminating the filaments, a prime source of trouble is also avoided because the filaments generally have a limited life expectancy. The absence of filaments also means that a warm-up period is not required before the device can operate properly. In other words, the solid-state component operates the instant it receives electrical power.

Solid state components are also able to operate with very low voltages (between 1 and 25 volts) while vacuum tubes usually require an operating voltage of 100 volts or more. This means that solid state components generally use less power than vacuum tubes and are, therefore, more suitable for use in portable equipment which obtains its power from batteries. The lower voltages are also much safer to work with. Pocket-size radios, hand held calculators, and small battery operated television receivers are typical examples of devices which take advantage of highly efficient, power saving components.

The small size of the solid-state component also makes it suitable for use in portable electronic equipment. Although equipment of this type can be constructed with vacuum tubes, such equipment would be much larger and heavier. A typical transistor is only a fraction of an inch high and wide while a vacuum tube of comparable performance may be an inch or more wide and several inches high. The small size also means a significant weight savings.

Solid-state components are much less expensive than comparable vacuum tube components. The very nature of a solid-state component makes it suitable for production in mass quantities which reduces cost. In fact, large numbers of solid-state components can be constructed as easily and quickly as a single component.

The most sophisticated semiconductor devices are integrated circuits. These are complete circuits where all of the components are constructed with semiconductor materials in a single microminiature package. These devices not only replace individual electronic circuits but also complete pieces of equipment or entire systems. Entire computers and radio receivers can be constructed as a single device no larger than a typical transistor. Integrated circuits have taken us one step farther in improving elec-

tronic equipment through the use of semiconductor materials. All electronic equipment has benefited from solid state components and particularly from the development of integrated circuits.

Disadvantages

Although solid-state components have many advantages over the vacuum tubes that were once widely used, they also have several inherent disadvantages. First, solid-state components are highly susceptible to changes in temperature and can be damaged if they are operated at extremely high temperatures. Additional components are often required simply for the purpose of stabilizing solid-state circuits so that they will operate over a wide temperature range. Solid-state components may be easily damaged by exceeding their power dissipation limits and they may also be occasionally damaged when their normal operating voltages are reversed. In comparison, vacuum tube components are not nearly as sensitive to temperature changes or improper operating voltages.

There are still a few areas where semiconductor devices cannot replace tubes. This is particularly true in high power, ultra high radio frequency applications. However, as semiconductor technology develops, these limitations are gradually being overcome.

Despite the several disadvantages just mentioned, solid-state components are still the most efficient and reliable devices to be found. They are used in all new equipment designs and new applications are constantly being found for these devices in the military, industrial, and consumer fields. The continued use of semiconductor materials to construct new and better solid-state components is almost assured because the techniques used are constantly being refined thus making it possible to obtain even superior components at less cost.

Semiconductors have had a profound effect on the design and application of electronic equipment. Not only have they greatly improved existing equipment and techniques by making them better and cheaper, but also they have permitted us to do things that were not previously possible. Semiconductors have revolutionized the electronic industry and they continue to show their even greater potential. Your work in electronics will always involve semiconductor devices. Examine this unit on semiconductor fundamentals carefully and you will benefit from the resulting knowledge.

1. Components which are made from semiconductor materials are often referred to as _electronic_ components.

2. Components made from semiconductor materials do not require a warm-up period.

   (A.) True
   B. False

3. Transistors are more reliable and have a longer life expectancy than tubes.

   (A.) True
   B. False

4. Components made from semiconductors require high operating voltages.

   A. True
   (B.) False

5. Practically all modern portable and compact electronic equipment utilizes _electronic_ components.

6. Semiconductor type components have almost completely replaced the older _vacum tubes_ components.

7. Components made from semiconductors are more susceptible to temperature changes than tubes.

   (A.) True
   B. False

8. By using semiconductor materials it is possible to construct entire circuits on a single chip.

   (A.) True
   B. False

9. List four advantages of semiconductor components over vacuum tubes.

   Size

   ruggedness

   Voltage Necessary to operate

   Cost

10. The three most common semiconductor components are _diodes_, _transistors_ and _Intergrated Circuits_

11. The primary function of a semiconductor component is to _Control_ the _Currents_ in a circuit.

12. The most advanced, complex and sophisticated solid-state device is the _Intergrated Circuits_.

Semiconductor Materials

The term semiconductor is used to describe any material that has characteristics which fall between those of insulators and conductors. In other words, a semiconductor will not pass current as readily as a conductor nor will it block current as effectively as an insulator. Some semiconductor materials are actually pure elements which are found in the periodic table of elements while other semiconductors may be classified as compounds. Typical examples of semiconductor materials that are natural elements are germanium (Ge) and silicon (Si).

The semiconductor elements that are suited to the greatest variety of electronic applications are germanium and silicon. Germanium is a brittle, grayish-white earth element that was discovered in 1886. This material may be recovered from the ash of certain types of coals in the form of germanium dioxide powder. This powder may then be reduced to pure germanium which is in solid form.

Silicon is a non-metallic element which was discovered in 1823. This material is found extensively in the earth's crust. A white or sometimes colorless compound known as silicon dioxide (also called silica) occurs abundantly in forms such as sand, quartz, agate, and flint. These silicon compounds can be chemically reduced to obtain pure silicon which is in a solid form. These two materials have atomic structures which may be easily altered to obtain specific electrical characteristics.

Once the pure material is available, it must then be suitably modified to give it the qualities necessary to construct a semiconductor device for a definite application.

13. Materials which have characteristics that fall midway between those of insulators and conductors are called _Semiconductors_

14. The two semiconductor materials most commonly used to manufacture electronic components are _gerAnium_ and _Silicon_.

15. The resistance of a semiconductor material compared to that of a good conductor like copper is:

A. higher.
B. lower.
C. about the same.

16. Semiconductor devices are made directly from pure semiconductor materials which have not been altered in any way.

A. True
B. False

Germanium And Silicon Atoms And Crystals

From your previous studies you know that metals such as copper and aluminum are used to carry current in an electrical circuit. You learned that these metals are classified as conductors because they offer minimum opposition to current flow. In previous studies you also learned that materials such as glass, rubber, and ceramic oppose the flow of electrical current and are therefore classified as insulators. You will now examine the atomic structure of two materials (germanium and silicon) which have characterisics that are between those of conductors and insulators and find out why these materials fall into a third classification known as semiconductors.

Definitions

Element: One of the known chemical materials that cannot be subdivided into simpler substances.

Atom: The smallest portion of an element that still exhibits all the characteristics of that element.

Semiconductor Atoms

Before we examine the structure of germanium and silicon atoms and crystals we must consider some important rules which pertain to the number and placement of the electrons that revolve around the nucleus of all atoms.

Atoms contain three basic components: protons, neutrons, and electrons. The protons and neutrons are located in the nucleus or center of the atom while the electrons revolve around the nucleus in orbits. The atom of each particular element will have a specific number of protons in its nucleus and an equal number of electrons, in orbit, if the atom is neutral (has no charge).

The exact manner in which the electrons are arranged around the nucleus is extremely important in determining the electrical characteristics of the element. Generally, each electron has its own orbit, but certain orbits are grouped together to produce what is referred to as a shell. For all of the elements that are known to exist, there can only be a maximum of seven shells.

The shell nearest the nucleus can only hold 2 electrons while the second shell can hold a maximum of 8 electrons. The third shell cannot hold more than 18 electrons and the fourth can hold no more than 32 and so on.

The outermost shell of a particular atom is called the valence shell and the electrons that orbit within this shell are referred to as valence electrons.

The arrangement of the protons and electrons in three different atoms are shown in Figure 1-1. Notice that the hydrogen atom has only one shell, while the carbon atom has two, and the copper atom has four. Also, notice that some of the shells contain less than the maximum number of electrons allowed.

In any particular atom, the outer shell can never hold more than 8 electrons. When exactly 8 electrons are present in the outer shell, the atom is considered to be completely stable and it will neither give up or accept electrons easily. Elements which have atoms of this type are neon and argon. These elements are classified as inert gases and they resist any sort of electrical or chemical activity.

When an atom has 5 or more electrons in its outer shell, it tries to fill its shell so that it can reach a stable condition. Elements of this type make good insulators because the individual atoms try to acquire electrons instead of giving them up. Therefore, the free movement of electrons from one atom to the next is inhibited. When an atom has less than 4 electrons in its outer shell it tends to give up these valence electrons easily. Elements which have atoms of this type make good conductors because they contain a large number of free-moving electrons which can randomly drift from one atom to the next.

When an atom contains exactly 4 electrons in its outer shell, it does not readily give up or accept electrons. Elements which contain atoms of this type do not make good insulators or conductors and are, therefore, referred to as semiconductors. The element carbon is a typical example of a semiconductor material. Notice that the carbon atom shown in Figure 1-1 has exactly 4 electrons in its outermost shell.

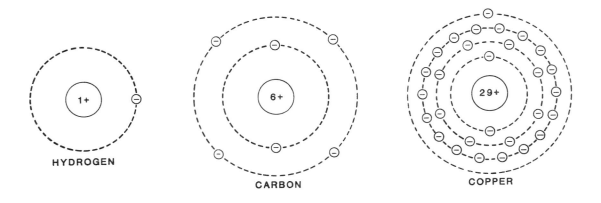

HYDROGEN   CARBON   COPPER

Figure 1-1 Diagram of typical Atoms.

The two semiconductor materials most commonly used in the construction of transistors and other types of related components are germanium and silicon. Both of these materials are made up of atoms which have 4 electrons in their outermost or valence shells. A single germanium atom is shown in Figure 1-2. Notice that the germanium atom has four shells and the distribution of the electrons from the first shell to the outer shell is 2, 8, 18 and 4. Therefore, a total of 32 electrons rotate around the nucleus of the atom and a total of 32 protons are contained within the nucleus.

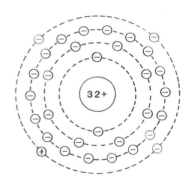

Figure 1-2 The Germanium Atom.

A single silicon atom is shown in Figure 1-3. Notice that this atom has only three shells and the distribution of the electrons from the first shell to the valence shell is 2, 8, and 4. The atom has a total of 14 electrons revolving around its nucleus and a total of 14 protons in its nucleus.

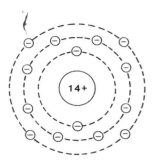

Figure 1-3 The Silicon Atom.

The important point to consider when examining the structure of germanium and silicon atoms is that both of these atoms have exactly 4 valence electrons. Although both of these atoms contain additional electrons, it is the valence electrons that determine the relative ease with which electrical current can flow through the germanium and silicon materials. To simplify the discussions and illustrations which follow, we will use diagrams of the germanium and silicon atoms which show only the four valence electrons surrounding a central core which consists of the nucleus and the inner shells. These simplified diagrams are shown in Figure 1-4. Notice that the core of the germanium atom is identified by the symbol Ge and the core of the silicon atom is identified by the symbol Si. In each case the nucleus is surrounded by four electrons.

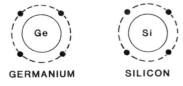

Figure 1-4 Simplified Germanium and Silicon Atoms.

Semiconductor Crystals

The individual atoms within a semiconductor material such as germanium are arranged as shown in Figure 1-5. Each atom shares its four valence electrons with four neighboring atoms as shown. This sharing of electrons creates a bond which holds the atoms together. This electron-pair bond is commonly referred to as a covalent bond, and it occurs because each of the atoms in the structure tries to take on additional electrons in order to fill its valence shell with eight electrons. The end result is a structure which has a lattice-like appearance and is often referred to as a crystal lattice.

Silicon atoms combine in the same manner as germanium atoms to form the same type of crystalline structure shown in Figure 1-5. The silicon atoms maintain covalent bonds just like the germanium atoms.

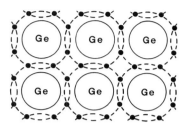

Figure 1-5 Simplified Diagram of Germanium Crystal Structure.

The germanium and silicon crystal lattices just described are free from impurities and therefore represent pure or ideal materials. Such crystals are often referred to as intrinsic materials. The construction of solid-state components such as transistors depends on the use of these pure or intrinsic semiconductor materials.

17. A neutral atom will have an equal number of ___Protons___ and ___nutrons___.

18. The electrons surrounding the nucleus of an atom are grouped into various ___Shells___.

19. There can never be more than ___Seven___ shells in any atom.

20. The shell closest to the nucleus of an atom cannot hold more than ___2___ electrons.

21. The outermost shell is referred to as the ___Valence___ shell.

22. The outermost shell can never hold more than ___8___ electrons.

23. A neon atom is completely stable because it has ___8 electrons in Valence Shell___

24. The electrical characteristics of an element are determined by the number of electrons in the ___Valence___ shell of its atoms.

25. Elements which are made up of atoms that have 4 electrons in their outer shells are called ___Semiconductors___.

26. The atoms in semiconductor materials are arranged in the form of a ___Crystal___ ___lattice___.

27. Each atom in a semiconductor material shares its electrons with four neighboring atoms to form a ___Crystal___ ___lattice___.

28. Intrinsic semiconductors do not contain ___Impurities___.

Conduction In Intrinsic Germanium And Silicon

Due to the crystalline structure of pure semiconductor materials such as germanium and silicon, each nucleus within the material sees eight valence electrons even though each atom actually has only four. Therefore, each atom tends to be stable and will not easily give up or accept electrons. However, this does not mean that pure semiconductors must, under all conditions, resist any sort of electrical activity in the same manner as the inert gases previously described. The reason for this is that another factor must be considered. This factor is temperature. The electrical characteristics of semiconductor materials are highly dependent upon temperature.

Low Temperature Characteristics

At extremely low temperatures, the valence electrons are held tightly to their parent atoms which are in covalent bonds and are not allowed to drift through the crystalline structures of either semiconductor material. Since the valence electrons cannot drift from one atom to the next, the material cannot support current flow at this time. Therefore, at extremely low temperatures, pure germanium and silicon crystals function as insulators.

High Temperature Characteristics

As the temperature of a germanium or sili-con crystal is increased, the valence electrons within the material become agitated and some of them will occasionally break away from the covalent bonds. Therefore, a small number of electrons will be free to drift from one atom to the next in a random manner. These free-moving electrons or free electrons are able to support a small amount of electrical current if a voltage is applied to the semiconductor material. In other words, as the temperature of the semiconductor material increases, the material begins to acquire the characteristics of a conductor. For all practical purposes, however, enough heat energy is available even at room temperature to produce a small number of free electrons which can support a small amount of current. Only when the semiconductor materials are exposed to extremely high temperatures, can a point be reached where they will conduct current as well as an ordinary conductor. Under normal conditions this high temperature usage of semiconductors is never encountered.

Holes

To understand exactly why a semiconductor is able to allow current to flow, we must take a closer look at the internal structure of the material. When an electron breaks away from a covalent bond, an open space or vacancy exists in the bond. The space that was previously occupied by the electron is generally referred to as a hole. A hole simply represents the absence of an electron. Since an electron has a negative charge, the hole represents the absence of

Doping- Adding Trivalent or Pentavalent
Tetravalent (4)
Intrinsic (pure)
Covelent Bonding
Crystal Lattice
Thermal Agitation
Electron-Hole pairs

a negative charge. This means that the hole has the characteristics of a positively charged particle. Each time an electron breaks away from a covalent bond, a hole is created. Each corresponding electron and hole is referred to as an electron-hole pair. A typical electron-hole pair is shown in Figure 1-6. Notice that the hole is represented by a plus sign to indicate positive. The electron is simply shown as a dot although it does have a negative charge. The semiconductor material shown could be either germanium or silicon as indicated.

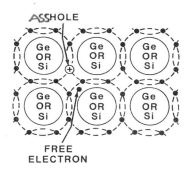

Figure 1-6 An Electron-Hole Pair in a Semiconductor Material.

The number of electron-hole pairs produced within a semiconductor material increases with temperature. Even at room temperature, a small number of electron-hole pairs exist. Some of the free electrons drift randomly. Holes absorb some of these electrons. This means that some electrons will simply jump from one shell to a shell which contains a hole. If an electron jumps from one shell to fill in a hole, another hole is created where the electron leaves the shell.

The hole, therefore, appears to move in the opposite direction of the electron. If another electron moves into the hole that was just created, another hole is produced and the previous hole appears to move randomly through a pure semiconductor material. Thus, the terms hole flow and electron flow.

Current Flow

When a pure semiconductor material such as a germanium or silicon is subjected to a difference-of-potential or voltage as shown in Figure 1-7, the negatively charged free electrons are attracted to the positive terminal of the voltage source. The positive holes that are created by the free electrons drift toward the negative terminal of the voltage source. As the free electrons flow into the positive terminal of the voltage source, an equal number of electrons leave the negative terminal of the voltage source. These electrons are injected into the left side of the

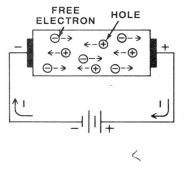

Figure 1-7 Current Flow in a Semiconductor Material.

semiconductor material where many of these electrons are captured or absorbed by holes. As the holes and electrons recombine in this manner, the holes cease to exist. Therefore the holes constantly drift to the left and then disappear while the electrons flow to the right where they are drawn out of the material and into the positive terminal of the voltage source.

It is important to remember that current flow in a semiconductor material consists of both electrons and holes. The holes function like positively charged particles while the electrons are actually negatively charged particles. The holes and electrons flow in opposite directions and the number of electron-hole pairs produced within a material increase as the temperature of the material increases. Since the amount of current flowing in a semiconductor is determined by the number of electron-hole pairs in the material, the ability of a semiconductor material to pass current increases as the temperature of the material increases.

It is also important to note that current flow in a semiconductor is somewhat different than current flow in a conductor. When we consider current flow in a semiconductor, we must consider the movement of holes as well as electrons. However, in a conductor we are concerned only with the number of free-electrons that are available.

Self-Test Review

29. At extremely low temperatures, a semiconductor is basically a/an _Insulator_.

30. At extremely high temperatures, a semiconductor will function as a/an _Conductor_.

31. Electrons which break away from covalent bonds and drift through a semiconductor are called _Free electrons_.

32. When an electron breaks away from a covalent bond, a hole is created.

    (A.) True
    B. False

33. An electron has a/an _Negative_ charge.

34. A hole has a/an _positive_ charge.

35. When a pure semiconductor is subjected to a voltage, free electrons flow toward the _positive_ terminal.

36. Holes move in the same direction as the electrons.

    A. True
    (B.) False

37. The number of electron-hole pairs within a semiconductor increases as temperature _Increases_.

38. The resistance of a semiconductor _decreases_ as the temperature decreases.

Conduction In Doped Germanium And Silicon

Pure semiconductor materials contain only a small number of electrons and holes at room temperature and therefore conduct very little current. However, the conductivity of these materials can be increased considerably by a process known as doping. Pure semiconductor materials such as germanium and silicon are doped by adding other materials, called impurities, to them when they are produced. Basically there are two types of impurities that are added to germanium and silicon crystals. One type of impurity is referred to as a pentavalent material because it is made up of atoms which have five valence electrons. The second type of impurity is referred to as a trivalent material because each of the atoms in this material has three valence electrons.

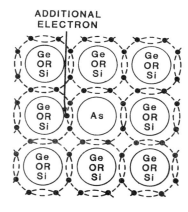

Figure 1-8 Semiconductor Material Doped with Arsenic.

N-Type Semiconductors

When a pure semiconductor material is doped with a pentavalent element, such as arsenic (As), some of the atoms in the crystal lattice structure of the tetravalent semiconductor are replaced by arsenic atoms. As a result, the crystal lattice of the semiconductor is like that shown in Figure 1-8. As shown in the figure, the arsenic atom has replaced one of the semiconductor atoms and is sharing four of its valence electrons with adjacent atoms in a covalent bond. However, the fifth electron is not part of a covalent bond and can be easily freed from the atom.

This arsenic atom is called a donor atom because it donates a free electron to the crystal lattice. Actually, there is a large number of donor atoms in the crystal lattice. Consequently, there are many free electrons in the semiconductor.

If a voltage is applied to an N-type semiconductor as shown in Figure 1-9, the free electrons contributed by the donor atoms will flow toward the positive terminal of the battery. However, some additional free electrons will also flow toward the positive terminal. These additional free electrons are produced as electrons break away from their covalent bonds to create electron-holepairs. This is identical to the action which takes place in a pure semiconductor material. The corresponding holes which are produced are then moved toward the negative terminal.

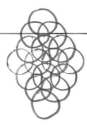

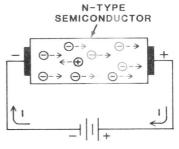

Figure 1-9 Conduction in an N-Type Semiconductor.

At normal room temperature the number of free electrons provided by the donor atoms will greatly exceed the number of holes and electrons that are produced by the breaking of covalent bonds. This means that the number of electrons flowing in the N-type semiconductor will greatly exceed the number of holes. The electrons, being in the majority, are therefore referred to as the majority carriers while the holes, which are in the minority, are referred to as the minority carriers in N-type material.

P-Type Semiconductors

Doping pure semiconductor material with a trivalent element, such as gallium (Ga), causes some of the semiconductor tetravalent atoms to be displaced by trivalent atoms. As shown in Figure 1-10, each trivalent atom shares its valence atoms with three adjacent atoms in the semiconductor

crystal lattice structure. However, the fourth tetravalent atom does not share a covalent bond because of the missing electron. This results in a hole in the crystal lattice. A large number of holes are present in the semiconductor because many trivalent atoms have been added. These holes readily accept electrons from other atoms. However, when a given hole is filled by an electron from another atom, the electron leaves another hole. Therefore, the holes drift from one covalent bond to another in the direction opposite to that of electron movement. Consequently, holes behave like positively charged particles. The acceptor atoms remain fixed within the crystal lattice even though the holes can move freely.

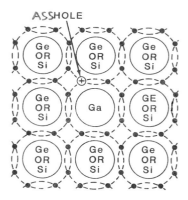

Figure 1-10 Semiconductor Material Doped with Gallium.

If a voltage is applied to a P-type semiconductor as shown in Figure 1-11, the holes provided by the acceptor atoms move from the positive to the negative terminal. These holes move in the same manner previously described. In other words, as each electron moves into a hole, another hole is created. Since electrons move toward the positive terminal, the holes move in the opposite direction or toward the negative terminal.

In addition to the holes provided by the acceptor atoms, many additional holes are also found in the P-type semiconductor material. These holes are produced as electrons break away from covalent bonds to create electron-hole pairs and are, therefore, produced in the same manner as the holes which appear in a pure semiconductor. These additional holes are also attracted toward the negative terminal while the corresponding electrons that are produced are attracted toward the positive terminal.

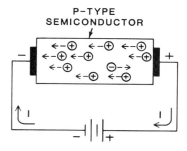

Figure 1-11 Conduction in a P-type Semiconductor.

Under normal conditions, the number of holes provided by the acceptor atoms will greatly exceed the number of holes and electrons that are produced by the breaking of covalent bonds. The number of holes flowing in the P-type semiconductor will therefore, greatly exceed the number of free electrons in the material. The holes being in the majority are referred to as the majority carriers and the electrons, which are in the minority, are referred to as the minority carriers in P-type material.

It is important to note that the semiconductor materials just described are referred to as N-type and P-type semiconductors because the majority carriers within these materials are electrons and holes which respectively have negative and positive charges. The N-type and P-type materials themselves are not charged. In fact, both materials are electrically neutral or uncharged. This is because each atom within each type of material has an equal number of electrons and protons. Also, the movement of holes and electrons does not cause the materials to become charged.

It is also important to understand that N-type and P-type semiconductors have a much higher conductivity than pure semiconductors. Also, the conductivity of these materials can be increased or decreased by simply adding more or less impurities. In other words, the more heavily a semiconductor is doped, the lower its electrical resistance.

39. The conductivity of a semiconductor can be increased by a process known as _doping_.

40. A pure semiconductor is doped by adding _impurities_ .

41. An impurity material which is made up of atoms that have five valence electrons is called a _pentavalent_ material.

42. An impurity material which is made up of atoms that have three valence electrons is called a _trivalent_ material.

43. Donor atoms add _Electrons_ to semiconductor crystals.

44. Acceptor atoms create _Holes_ in semiconductor crystals.

45. In an N-type semiconductor the majority carriers are _electrons_ .

46. In a P-type semiconductor the majority carriers are _protons_ .

47. Increased doping causes the electrical resistance of a semiconductor to _lower_ .

Semiconductors are substances that are both poor conductors and poor insulators. The most important semiconductors are the chemical elements that have four electrons in the outermost, or valence, shell of their atoms. The two semiconductor materials used in the manufacture of solid-state components are the elements silicon and germanium; the use of silicon is more common.

It is a characteristic of tetravalent atoms to readily share valence electrons with other atoms. This sharing of valence atoms is called a covalent bond. These covalent bonds cause the atoms to form a crystal lattice.

Pure (intrinsic) semiconductors function as insulators at low temperatures because valence electrons are held tightly in their shells. However, as temperature increases, valence electrons are occasionally able to break their covalent bonds and become free electrons, leaving holes in the crystal lattice. If a voltage is applied under these conditions, electrons can move in one direction while holes move in the opposite direction, allowing the semiconductor to act as a fairly good conductor.

Intrinsic semiconductors have limited use in electronics. However, they can be modified, by doping, to meet specific electrical requirements. Doping is the deliberate addition of an impurity having the desired characteristics to the semiconductor element.

Doping a semiconductor with a pentavalent element, such as arsenic, adds a large number of free electrons to the crystal lattice. These electrons freely move from atom to atom in one direction within the crystal lattice when a voltage is applied.

Hole movement is also present in the crystal lattice, but the quantity of holes is much smaller than the quantity of electrons. For this reason electrons are the majority carriers of current and holes are the minority carriers in a semiconductor that is doped with pentavalent atoms. Semiconductors that are doped with pentavalent atoms are N-type semiconductors.

Doping a semiconductor with a trivalent element, such as gallium produces a P-type semiconductor. The valence electrons in the outer shell of each trivalent atom form covalent bonds with three of the four adjacent tetravalent atoms in the crystal lattice. However, the absence of a fourth valence electron in the trivalent atom causes a hole in the crystal lattice. The large number of trivalent atoms in the crystal lattice causes a large number of holes; each hole is capable of accepting a free electron. However, the number of holes greatly outnumbers the free electrons. Consequently, holes are the majority current carriers and electrons are the minority carriers in P-type semiconductors.

Unit 2
Semiconductor Diodes

Contents

Introduction

In the previous Unit you examined the semiconductor materials that are used to construct various types of solid-state components. Now you will learn how these materials are used to construct one of the most important types of solid-state components, the semiconductor diode. Although simple in construction and operation, semiconductor diodes are widely used in many types of electronic equipment. Rectification, waveshaping, circuit protection, and logic operations are only a few of its diversified applications. In addition, there are a variety of diode types each of which is optimized for a particular job. In this Unit you will learn semiconductor diode fundamentals. In later units you will learn about the many different diode types and their applications.

Since transistors, integrated circuits, and other solid-state components are constructed in basically the same manner as diodes, an understanding of diode construction and operations is an essential first-step in understanding semiconductor devices.

Read the unit objectives carefully to ensure that you understand what is to be learned in this unit. This will help you relate the material to the order in which it is presented. Reading the objectives will also provide an insight into which text material is required for a better understanding of semiconductor diodes.

Unit Objectives

When you have completed this unit you should be able to:

1. Describe the construction of a semiconductor diode's PN junction.

2. Recognize the schematic symbol of a semiconductor diode.

3. Explain the difference between an atom and an ion.

4. Describe the electrical characteristics of a PN junction.

5. Identify the two parts of a diode on the schematic symbol and on an actual diode.

6. Describe the effects of forward and reverse bias on the basic of a junction diode.

7. Interpret a graph of diode voltage-current characteristics.

8. Use an ohmmeter to determine if a diode PN junction is operational or faulty.

9. List at least four applications for semiconductor diodes.

10. Identify at least four different diode packages.

As you learned in the previous Unit, a pure or intrinsic semiconductor can be doped with a pentavalent or trivalent material to obtain two basic types of semiconductor materials. These doped semiconductors are referred to as N-type or P-type materials because the majority carriers within them are either electrons or holes respectively. The N-type and P-type designations do not imply that the materials have negative and positive charges. In fact, both types are electrically neutral because the atoms within each semiconductor, including the impurity atoms, contribute an equal number of protons and electrons to the respective materials. Even though some of the electrons and holes are free to move around in a random manner, the semiconductor materials do not acquire an overall electrical charge.

Ions

Even though N-type and P-type semiconductors are electrically neutral, within each type of semiconductor material, independent electrial charges do exist. For example, each time an atom gives up an electron, it loses a negative charge and becomes electrically unbalanced. In other words, the positive protons, in the nucleus of the atom outnumber the electrons revolving around the nucleus and the atom takes on a positive charge. Also when a neutral atom takes on a additional electron, it has more negative electrons than positive protons and therefore assumes a negative charge. These electrically charged atoms are referred to as ions. A positively charged atom is therefore

referred to as a positive ion and a negatively charged atom is referred to as a negative ion.

The free electrons and holes which drift throughout the semiconductors also possess negative and positive charges respectively. However, since these charges are able to move, they are referred to as mobile charges. Within a N-type or P-type semiconductor, an equal number of mobile charges and ionic charges will exist and since these charges are equal and opposite, the semiconductor material remains electrically neutral.

The internal structure of N-type and P-type semiconductors may be illustrated in a simplified manner as shown in Figure 2-1. The N-type semiconductor is doped with a pentavalent impurity and therefore contains many donor atoms which contribute free electrons that can drift through the material. The donor atoms take on positive charges and become positive ions when they release free electrons. These are represented by plus signs surrounded by small circles as shown. The free electrons which accompany the donor atoms are represented by minus signs.

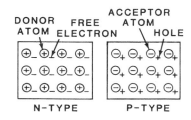

Figure 2-1 Doped semiconductors.

The P-type semiconductor is doped with a trivalent impurity and therefore contains many acceptor atoms. These atoms easily accept or absorb electrons from the semiconductor material and become negatively charged atoms or negative ions. The negatively charged acceptor atoms are represented by minus signs that are surrounded by small circles as shown. The holes created by these atoms act like positively charged particles and are represented by plus signs as indicated. By representing the doped semiconductors in this manner it becomes relatively easy to analyze the action which takes place when they are combined to form diodes and other types of solid-state components.

Junction Diodes

Now let's consider the action which takes place when doped semiconductors are combined to form a diode. Basically a diode is created by joining N-type and P-type semiconductors as shown in Figure 2-2. When these oppositely doped materials come in contact with each other, a junction is formed where they meet. Such diodes are referred to as junction diodes and can be made of either silicon or germanium. When the junction is formed, a unique action takes place.

Depletion Region

The mobile charges, free electrons and holes, in the vicinity of the junction are strongly attracted to each other and there-fore drift toward the junction. The accumulated charges at each side of the junction serve to increase the attraction even more. Eventually some of the free electrons move across the junction and fill some of the holes in the P-type material. As the free electrons cross the junction, the N-type material becomes depleted of electrons in the vicinity of the junction. At the same time, the holes within the P-type material become filled and no longer exist. This means that the P-type material also becomes depleted of holes near the junction. This region near the junction where the electrons and holes become depleted is referred to as the depletion region. The depletion region extends for only a short distance on each side of the junction as shown in Figure 2-2.

It is important to remember that free electrons and holes are actually the majority carriers for the N-type and P-type materials. Therefore, no majority carriers exist within the depletion region. Also, it is important to note that the N-type and P-type materials are no longer neutral or uncharged. In other words the N-type material has lost free electrons, thus causing the positive donor atoms (ions) to outnumber the free electrons. The N-type material therefore takes on a positive charge near the junction. The P-type material has lost holes which means that the negatively charged acceptor atoms (ions) will outnumber the holes. The P-type material therefore, takes on a negative charge near the junction. This means that opposite charges now exist on each side of the junction.

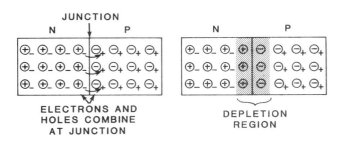

Figure 2-2 Characteristics of a PN Junction.

The depletion region does not continue to become larger and larger until the N-type and P-type materials are completely depleted of majority carriers. Instead, the action of the electrons and holes combining at the junction tapers off very quickly. Therefore, the depletion area remains relatively small. The size of the depletion region is limited by the opposite charges which build up on each side of the junction. The negative charge which accumulates in the P-type material eventually becomes great enough to repel the free electrons and prevent them from crossing the junction. The positive charge which accumulates in the N-type material also helps to stop the flow of free electrons by attracting and holding them so that they cannot move across the junction.

Barrier Voltage

The opposite charges that build up on each side of the junction create a difference in potential or voltage. This difference in potential effectively limits the size of the de-

pletion region by preventing the further combination of electrons and holes. It is referred to as the potential barrier or the barrier voltage. Although this barrier voltage exists across the PN junction, its effect can be represented by an external battery as shown in Figure 2-3.

The barrier voltage produced within a PN junction will usually be on the order of several tenths of a volt. This voltage will always be higher for silicon PN junctions than for germanium PN junctions. For example, a PN junction made from doped germanium will have a typical barrier voltage of 0.3 volts while a PN junction made from doped silicon will have a typical barrier voltage of 0.7 volts. The barrier voltage exists inside the junction diode and therefore cannot be measured directly, but its presence becomes apparent when an external voltage is applied to the diode. Later in this unit you will see why this is true.

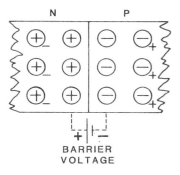

Figure 2-3 Barrier Voltage in a PN Junction

1. Electrically charged atoms are referred to as ___Ions___ .

2. The free electrons and holes which drift throughout a semiconductor are referred to as ___Mobil___ ___Charges___ .

3. A diode is formed by joining ___P___ and ___N___ semiconductor materials.

4. When a PN junction is formed, a ___depletion___ region is created in the area near the junction.

5. The opposite charges which build up on each side of a PN junction create a difference of potential which is known as the:

    A. mobile charge
    B. bias voltage
    C. barrier voltage

6. The barrier voltage within a diode will usually be in which of the following ranges?

    A. 10 to 100 mV
    B. 100 mV to 1 V
    C. 1 V to 10 V
    D. 10 V to 100 V

7. A voltmeter may be used to directly measure the barrier voltage within a diode.

    A. True
    B. False

8. An understanding of diode operation can only be obtained by examining the action which takes place at the diode's: ___PN Junction___ .

Diode Biasing

Whenever diodes are used in electrical or electronic circuits, they are subjected to various voltages and currents. The polarities and amplitudes of the voltages and currents must be such that proper diode action takes place. We generally refer to the voltages applied to a semiconductor diode as bias voltages. Let's see how these voltages affect and control the diode's operation.

Forward Bias

In the previous discussion, you saw that free electrons and holes, the majority carriers, combine at the junction to produce a depletion region. The depletion region represents an area that is void of majority carriers but at the same time contains a number of positively charged donor atoms (ions) and negatively charged acceptor atoms (ions). These positive and negative charges are separated at the junction and effectively create a barrier voltage which opposes any further combination of majority carriers. It is important to remember that this action takes place in a PN junction diode that is not subjected to any external voltage.

When a PN junction diode is subjected to a sufficiently high external voltage V as shown in Figure 2-4, the device will function in a somewhat different manner. Notice that the negative and positive terminals of a battery are connected to the N and P sections of the diode respectively. An external resistor R is used to limit the current level to a safe value. Under these conditions the free electrons in the N section are repelled

by the negative battery terminal and forced toward the junction where they will neutralize the positively charged donor atoms (positive ions) in the depletion region. During this same period of time, the free electrons that had initially accumulated to create a negative charge on the P side of the junction are attracted toward the positive battery terminal. Therefore the negative charge on the P side of the junction is also neutralized. This means that the positive and negative charges which form the internal barrier voltage are effectively neutralized and no barrier voltage will be present to stop the combining of majority carriers at the junction. The PN junction diode is therefore able to support a continuous flow of current at this time. This action will occur only if the battery voltage is greater than the barrier potential.

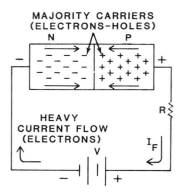

Figure 2-4 Forward-biased PN Junction Diode.

Since the diode is now subjected to an external voltage, a constant supply of electrons flow into the N section of the diode. These electrons drift through the N-type material toward the junction. The movement of these electrons through the N section is sustained by the free electrons (majority carriers) that exist within this material. At the same time, the holes (majority carriers) in the P section also drift toward the junction. The electrons and holes combine at the junction and effectively disappear as they neutralize each other. However, as these electrons and holes combine and are effectively eliminated as charge carriers, new electrons and holes appear at the outer edges of the N and P sections. The majority carriers therefore continue to move toward the junction as long as the external voltage is applied. This action is shown in Figure 2-4.

It is convenient to analyze the action which takes place in the P section of the diode by considering the movement of holes instead of electrons. However, it is important to realize that electrons do actually flow through the P-type material. The electrons are attracted by the positive terminal of the battery and as the electrons leave the P section and enter the battery, holes are created at the outer edge of the P section. These holes drift toward the junction where they combine with electrons and effectively disappear.

The important point to note at this time is that electrons do flow through the entire PN junction diode when it is subjected to a sufficiently high external voltage. At this time the diode is said to be conducting current in the forward direction. Also, the diode is considered to be forward-biased by the external voltage. The current which flows through the forward-biased diode shown in Figure 2-4 is limited by the resistance of the P-type and N-type semiconductor materials as well as the external resistance R. Normally the diode resistance is quite low. Connecting a forward bias voltage directly to the diode will result in a current large enough to generate sufficient heat to destroy the diode. For this reason, forward-biased diodes are usually connected in series with a resistor or some other device which will limit the current to a safe level.

A forward-biased diode will conduct current as long as the external bias voltage is sufficiently high and the polarity is correct. For example, if the diode is constructed from germanium, a forward bias of approximately 0.3 volts will be required before the diode can begin to conduct. Silicon diodes require a forward bias of approximately 0.7 volts in order to begin conducting. The external voltage applied to the diode must be large enough to neutralize the depletion area and therefore neutralize the barrier voltage that exists across the PN junction of the diode. Once this internal voltage is overcome, the diode will conduct in the forward direction.

The polarity of the external DC bias voltage must also be correct with respect to the P and N sections of the diode. The negative terminal of the bias source must be connected to the N section and the positive terminal should be connected to the P section to achieve the forward-biased condition.

Once the diode is conducting a voltage will be dropped across the device. This occurs because the diode's semiconductor material has a low but finite resistance value and the current flowing through it must produce a corresponding voltage drop. As it turns out, this forward bias voltage drop is approximately equal to the barrier potential. This is 0.3 volts for a germanium diode and 0.7 volts for a silicon diode.

The amount of forward bias current I_F is a function of the applied DC bias V, the forward voltage drop V_F, and the external resistance R. The relationship simply involves Ohm's law as indicated below:

$$I_F = \frac{V - V_F}{R}$$

For example, the forward current in a silicon diode with a bias voltage of 10 volts and an external resistor of 100 ohms is:

$$I_F = \frac{10 - .7}{100 \, \Omega}$$

$$= \frac{9.3}{100 \, \Omega}$$

$$= .093 \text{ amps or } 93 \text{ mA}$$

Reverse Bias

A forward-biased diode is able to conduct current in the forward direction because the external bias voltage forces the majority carriers together so that they can combine at the junction of the diode and create a continuous flow of current. In order to achieve this condition, the negative terminal of the battery is connected to the N section of the diode and the positive terminal is connected to the P section. However, if the battery connections are reversed as shown in Figure 2-5, the diode will operate in a different manner. The negative terminal of the battery is now connected to the P section of the diode while the positive terminal is connected to the N section. The diode is now considered to be reversed-biased.

Under these conditions the free electrons in the N section will be attracted toward the positive battery terminal, thus leaving a relatively large number of positively charged donor atoms (positive ions) in the vicinity of the junction as shown in Figure 2-5. In fact, the number of positive ions in the N section at times will even outnumber the positive ions that exist in an unbiased diode. This effectively increases the width of the depletion region on the N side of the junction causing the positive charge on the other side of the junction to increase.

At the same time, a number of electrons leave the negative terminal of the battery and enter the P section of the diode. These electrons fill the holes near the junction thus causing the holes to appear to move toward the negative terminal. A large number of negatively charged acceptor atoms (negative ions) are therefore created near the junction. This effectively increases the width of the depletion region on the P side of the junction.

The overall depletion region of the diode shown in Figure 2-5 is wider than the depletion region in the unbiased diode shown in Figure 2-2. This means that the opposite charges on each side of the junction are also larger and therefore create a higher barrier voltage across the junction. These opposite charges will build up until the internal barrier voltage is equal and opposite to the external battery voltage. Under these conditions the holes and electrons (majority carriers) cannot support current flow and the diode effectively stops conducting.

Actually, an extremely small current will flow through the diode shown in Figure 2-5. This small current is sometimes referred to as a leakage current or a reverse current and is designated as IR. It exists because of the minority carriers which are contained within the N and P sections of material and electrons in the P-type material. When the diode is reverse-biased as shown in Figure 2-5 the minority carriers are forced toward the junction where they combine and thus support an extremely small current. This action closely resembles the action which takes place in the forward-biased diode shown in Figure 2-4 but it is on a much smaller scale.

The number of minority carriers in the N and P materials is extremely small at room temperature. However, as temperature increases, a greater number of electron-hole pairs are generated within the two materials. This causes an increase in minority carriers and a corresponding increase in leakage current.

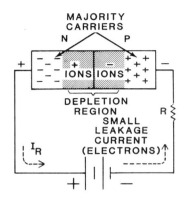

Figure 2-5 Reversed-biased PN Junction Diode

All PN junction diodes produce some leakage current when they are reversed-biased but this current is very small. It is only microamperes in germanium diodes and nanoamperes in silicon diodes. These currents are usually many orders of magnitude less than the usual forward current. It is important to remember that germanium diodes normally produce a higher leakage current than silicon diodes. Germanium devices are also more temperature sensitive than silicon devices. This disadvantage of germanium diodes is often offset by the desirable lower barrier potential and forward voltage drop.

We can sum up the operation of a PN junction diode in this manner. The diode is a unidirectional electrical device since it conducts current in only one direction. When it is forward biased, current flows through it freely since it acts as a very low resistance. When the diode is reverse biased, current does not flow through it. It simply acts as an open circuit or extremely high resistance.

Only a small, temperature sensitive leakage current flows in the reverse biased condition. The diode is effectively a polarity sensitive electrical switch. When forward biased the diode switch is closed. When reverse biased it is open.

9. The depletion region is neutralized when a diode is _Forward biased_

10. A diode is forward-biased when the negative terminal of the external battery is connected to the _Cathode (N)_ section of the diode.

11. A forward-biased diode acts like a switch that is _ⓐ closed_ .

12. The current flowing in a forward-biased diode is sustained by the _majority_ carriers in the N and P materials.

13. The amount of forward-bias voltage applied to a diode before it can conduct must be greater than the _barrier voltage_ .

14. A reverse-biased diode acts like a switch that is _open_ .

15. A diode is reverse-biased when the negative terminal of the external battery is connected to the _Anode (P)_ section of the diode.

16. When a diode is reverse-biased, its depletion area is _larger_ than when it is unbiased.

17. The current that flows through a reverse-biased diode is sustained by the _minority_ carriers in the N and P materials.

18. The barrier voltage produced within a reverse-biased diode is equal and opposite to the external voltage applied.

    A. True
    B. False

19. The small reverse current that flows through a reverse-biased diode is sometimes called a _____ current.

20. As temperature increases, the reverse current in a diode _____ .

21. An external source of 5 volts DC is applied to a germanium diode in series with a 1k ohm resistor. The positive source lead is connected to the P section of the diode. How much current flows, if any? _____

Diode Characteristics

Now that you have seen how a basic PN junction diode operates, you are ready to examine some of the important electrical characteristics of these devices. Since the characteristics of diodes vary considerably when they are subjected to various voltages and temperatures, it is usually best to plot the desired characteristics in a graphical manner. This makes it possible to analyze the operation of the device at a variety of points where the voltages, currents, or temperatures involved have specific and related values.

The graph in Figure 2-6 shows the amount of forward current and reverse current that flows through a typical PN junction diode when the device is first forward-biased and then reverse-biased. The diode's forward and reverse bias voltages, V_C and V_R, are plotted to the right and to the left respectively on the horizontal axis of the graph. The diode's forward and reverse currents, I_F and I_R, are plotted above and below the horizontal axis respectively to form the vertical axis of the graph. The point where the horizontal axis crosses the vertical axis is often referred to as the origin of the graph. This origin serves as a zero reference point for the four quantities involved. A graph like the one shown in Figure 2-6 is created by actually subjecting a diode to various forward and reverse voltages while measuring the current through the diode. However, certain precautions must be taken to insure that the diode is not damaged by excessive current or voltage.

When a large number of corresponding voltage and current values are plotted, a continuous curve is obtained as shown. For this reason, a graph like the one shown in Figure 2-6 is often referred to as a voltage-current or V-I characteristic curve. If you examine Figure 2-6 closely you will find that the forward current and voltages and the reverse currents and voltages are plotted to different scales. This is because the forward characteristics involve low voltages and high currents while the reverse characteristics involve relatively high voltages and low currents.

Germanium Diode

The V-I characteristic curve in Figure 2-6 is for a germanium diode. Let's consider its operation in detail.

Forward Characteristics

Figure 2-6 shows that the forward current through a germanium diode is extremely small and almost insignificant until the forward bias voltage across the diode increases beyond the value of 0.2 volts. Then the forward current increases as the forward bias voltage is increased still further. The increase in forward current really starts to occur as the external bias voltage overcomes the diode's internal barrier voltage. Once the bias voltage exceeds the barrier voltage (0.3 volts), the forward current increases very rapidly and at a linear rate because the diode then acts as a low resistance. If this forward current continued to rise, the diode would eventually be damaged by an excessive flow of current. Throughout the linear portion of the curve, the voltage across the diode is only several tenths of a volt as shown. While the forward voltage drop is not constant, it changes very little over a wide current range. A tremendous change in forward current occurs while the voltage across the diode changes only a small amount.

The point at which the bias voltage equals the barrier voltage is indicated in Figure 2-6.

Notice that this point occurs when the bias voltage is equal to 0.3 volts. Also notice that the diode's forward current is equal to 1 milliampere at this time and that this current can increase above 5 milliamperes while the corresponding voltage across the diode remains below 0.4 volts. Figure 2-6 therefore shows that the diode's internal barrier voltage is approximately 0.3 volts. However, it is important to realize that this voltage will vary slightly from one germanium diode to the next.

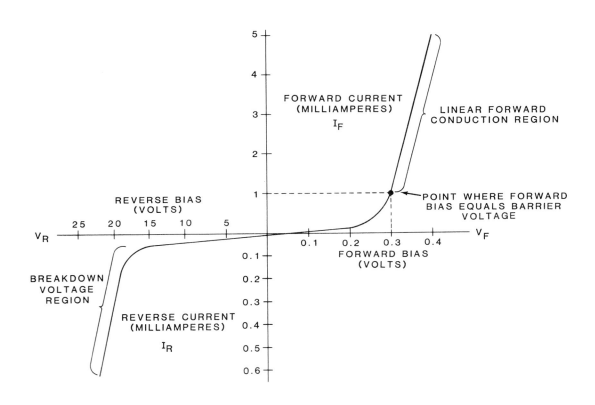

Figure 2-6 Typical Germanium Diode V-I Characteristics.

Reverse Characteristics

The V-I curve in Figure 2-6 also shows that when the diode is reverse-biased, the reverse current that flows is extremely small. Notice that the reverse current increases slightly as the reverse voltage increases but remains less than 0.1 milliamperes (100 microamperes) until the reverse voltage approaches a value of 20 volts. Then the reverse current suddenly increases to a much higher value. This sudden increase in reverse current results because the reverse bias voltage becomes strong enough to tear many valence electrons from their parent atoms and therefore increases the number of electron-hole pairs in the N and P materials. This causes an increase in minority carriers which in turn support a higher reverse current. In other words, the junction breaks down when the reverse bias voltage approaches 20 volts.

The voltage at which the sudden change occurs is commonly referred to as the breakdown voltage. This breakdown voltage will vary from one diode to the next since it is determined by the exact manner in which the diode is constructed. In certain cases, ordinary germanium diodes can be damaged when breakdown occurs; however, there are special diodes which are designed to operate in this region. These special devices, known as zener diodes, will be described in detail in the next Unit. When breakdown occurs, the diode no longer offers a high resistance to the flow of reverse current and therefore cannot effectively block current in the reverse direction. For these reasons, operation in the breakdown region is avoided when an ordinary PN junction diode is being used.

Silicon Diode

While a silicon diode operates the same as the germanium diode, there are some important differences in their characteristic curves. Let's look at these differences in detail.

Forward Characteristics

The V-I curve in Figure 2-7 shows the characteristics of a typical silicon diode. Notice that the forward characteristics of this diode are basically similar to those of the germanium diode previously described; however, there is an important exception. The internal barrier voltage of the silicon diode is not overcome until the forward bias voltage is equal to approximately 0.7 volts as shown. Beyond this point the forward current increases rapidly and at a linear rate. The corresponding forward voltage across the diode increases only slightly. The exact amount of forward voltage required to overcome the barrier voltage will vary from one silicon diode to the next but will usually be close to the 0.7 volts indicated in Figure 2-7.

Reverse Characteristics

The reverse characteristics of the silicon diode are also similar to those of the typical germanium diode previously described. However, the silicon diode has a much lower reverse current than the germanium type as indicated in Figure 2-7. Notice that the reverse current remains well below 0.1 milliamperes (100 microamperes) until the breakdown voltage of the device is reached. Then, as with the germanium unit, a rela-

tively high reverse current is allowed to flow. A breakdown voltage of 45 volts is indicated in Figure 2-7, however, this voltage will vary from one silicon diode to the next. Also, the reverse currents in many silicon diodes may be in the extremely low nano-ampere range, and therefore insignificant, for most practical applications.

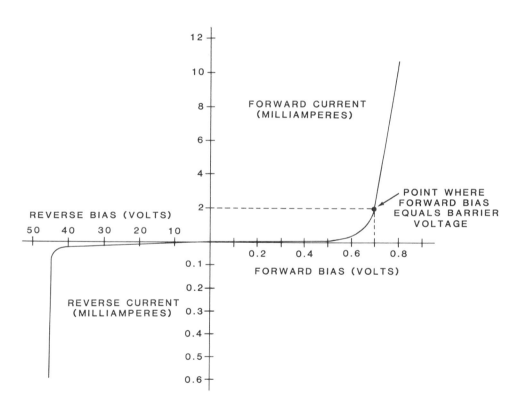

Figure 2-7 Typical Silicon Diode V-I Characteristics.

Diode Ratings

When the important characteristics of silicon and germanium diodes are compared it becomes apparent that either type can be damaged by excessive forward current. For this reason, manufacturers of these diodes usually specify the maximum forward current (IF max) that each type can safely handle. Also, both types can be damaged by excessive reverse voltages which cause the diode to breakdown and conduct a relatively large reverse current. To insure that the various diodes are not subjected to dangerously high reverse voltages, manufacturers of these devices usually specify the maximum safe reverse voltage that can be applied to each particular device. This maximum reverse voltage is commonly referred to as the peak inverse voltage, which is usually abbreviated PIV.

Temperature Considerations

In some critical applications it is also necessary to consider the effect that temperature has on diode operation. In general, the diode characteristic that is most adversely affected by changes in temperature is the diode's reverse current. This current is caused by the minority carriers that are present in the N and P sections of the diode as explained previously. At extremely low temperatures the reverse current through a typical diode will be practically zero. But at room temperature this current will be somewhat higher although still quite small. At extremely high temperatures an even higher reverse current will flow which in some cases might interfere with normal diode operation. These changes in reverse current as a result of temperature changes are shown in Figure 2-8. Notice that the

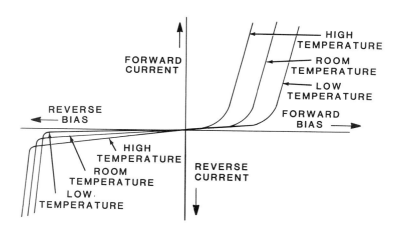

Figure 2-8 Relative changes in Current due to changes in Temperature.

breakdown voltage also tends to increase as temperature increases, however, this change is not great.

These same relative changes occur in both types of diodes, even though the reverse currents are generally higher in the germanium types. For both germanium and silicon diodes, the reverse or leakage current doubles for approximately every 10 degrees Centigrade rise in temperature.

The forward voltage drop across a conducting diode is also affected by temperature changes. This is illustrated in Figure 2-8. The forward voltage drop is inversely proportional to temperature. As the temperature rises, the voltage drop decreases. This effect is the same in both germanium and silicon devices.

flow. As you discovered earlier, the forward current must flow from the N section to the P section of the diode. This means that the forward current through the symbol must flow from the bar or rectangle to the triangle or arrow. In other words forward current flow is always against the arrow in the diode symbol. Also notice that the N and P sections of the diode and the corresponding portions of the diode symbol have been identified as the cathode and the anode respectively. These two terms were once widely used to identify the two principle elements within a vacuum tube diode. However, they are now commonly used to describe the two sections of a junction diode. The cathode (N-type) is simply the section of the diode that supplies the electrons and the anode (P-type) is the section that collects the electrons.

Diode Symbols

When diodes are shown in a circuit drawing or schematic, it is convenient to represent them with an appropriate symbol. The symbol most commonly used to represent the diode is shown in Figure 2-9 along with a typical PN junction diode. Notice that the P section of the diode is represented by a triangle (also called an arrow) and the N section is represented by a bar (also called a rectangle). The arrows that are placed beside the diode and its symbol indicate the direction of forward current (I_F) or electron

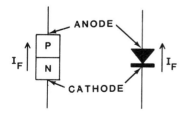

Figure 2-9 A typical Junction Diode and its Symbol

Figure 2-10 shows how forward-biased and reverse-biased diodes are represented in schematic form. Notice that when the negative and positive terminals of the battery are connected to the cathode and anode of the diode respectively, the diode is forward-biased and will conduct a relatively high forward current. (I_F). The resistor is added in series with the diode to limit this forward current to a safe value. Also notice that when the battery terminals are reversed, the diode is reverse-biased and only a very low reverse current (I_R) will flow through the device.

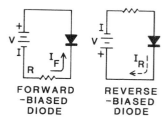

Figure 2-10 Forward and Reverse-biased diode circuits.

Desk-Top Experiment 1
Characteristics of Diodes

Introduction

This experiment will provide the experience necessary for you to analyze the difference between an ideal diode and the actual diode circuit.

Objectives

1. Calculate ideal load currents for a given circuit.

2. Calculate actual load currents for a given circuit.

3. Calculate the % of source voltage developed across the load in a given circuit.

4. Explain the practical limits placed on the value of R_L and the source voltage in a given circuit.

Procedure

On a separate sheet of paper draw the equivalent circuits for the procedures listed. You will then calculate the ideal and actual values of load current I_L, as well as the simulated forward conducting resistance (R_F) of the diode. Finally, you will determine the percentage of source voltage that is developed across the load resistor R_L.

1. Assume the diode in Figure 2-11 is an ideal diode. Calculate the load current, I_L.

$$I_L \text{ (ideal)} = \underline{\hspace{2cm}}.$$

2. Assume that the diode is a silicon type. Calculate the actual load current.

$$I_L \text{ (actual)} = \underline{\hspace{2cm}}.$$

3. Calculate the diode's simulated value of forward conducting resistance, R_F.

$$R_F \text{(simulated)} = \underline{\hspace{2cm}}.$$

4. Calculate the percentage of source voltage developed across R_L.

% of voltage developed across
$$R_L = \underline{\hspace{2cm}}.$$

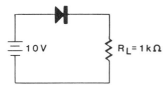

Figure 2-11 Circuit for Desk-Top Experiment 1

Discussion

The load current (I_L) for the circuit using an ideal diode assumes that the forward conducting resistance of the diode (R_F) is zero. Thus, the diode would be a perfect short when forward biased. However when the effect of the internal resistance of the conducting diode (R_F) is considered, the load current will decrease. Anytime a resistance is added in series, current decreases. In this case, the diode is a silicon type and the voltage drop across a silicon PN junction is considered to be 0.7 volts. The 0.7 volts is the voltage pressure required to keep current flowing through the barrier junction and is actually represented by a series opposing battery with a value of 0.7 volts.

Procedure (Cont.)

5. Assume that R_L is changed to 10 k ohms. Calculate the ideal load current.

    I_L (ideal) = _____.

6. Calculate the actual load current.

    I_L (actual) = _____.

7. Calculate the new simulated value of R_F.

    R_F = _____.

8. Calculate the percentage of source voltage developed across R_L.

    % of voltage developed across R_L = _____.

Discussion

You should have noticed that as R_L increased I_L decreased. The decreased I_L also flows through the diode and therefore, to maintain the 0.7 volt drop across the diode, the simulated value of R_F must have increased. Note also that the load current, I_L, changed by only 70 microamperes between the ideal and actual diode circuits.

Procedure (Cont.)

9. Change the source voltage to 100 volts and recalculate the ideal and actual load currents.

    I_L (ideal) = _____.

    I_L (actual) = _____.

10. Calculate the new simulated value of R_F.

    R_F (simulated) = _____.

11. Calculate the percentage of source voltage developed across R_L.

    % of source voltage developed across RL = _____.

Discussion

When the source voltage is increased to 100 volts, the load current increases but the voltage drop across RF remains at approximately 0.7 volts. Therefore, the simulated forward conducting resistance, RF of the diode must have decreased. The ratio of output voltage to source voltage is now over 99%. This means that more than 99% of the source voltage is now developed across the load.

Summary

If the source voltage is too low, the 0.7 volt drop across the internal resistance of the conducting silicon diode is a large portion of the total source voltage. When the load resistor, R_L, is low the load current is high and the simulated value of R_F is a much smaller value. Thus, the 0.7 volt drop across R_F remains essentially constant.

When the load resistor is large (100:1) when compared to the junction resistance, the junction resistance is not an important concern, provided the source voltage is high enough to offset the normal 0.7 volt drop across the PN junction.

As you can see by comparing your calculations, the simulated value of R_F can greatly affect the output voltage and load current. Therefore, to minimize the effect of R_F, it is important to consider both the amount of source voltage and the size of the load resistor.

When the source voltage is increased to 100 volts, the voltage dropped across the diode can usually be neglected.

The voltage drop across a solid state junction is actually not a constant, but it varies very little under most circuit conditions. The approximations of 0.3 volts for germanium junctions and 0.7 volts for silicon junctions are the standards selected for the Heathkit/Zenith educational courses.

In actual practice, the 0.3 volts for germanium and the 0.7 volts for silicon voltage drops across the junctions are almost constant. They are illustrated as a series opposing battery and a bulk resistance that is almost constant (for a given diode) under varing input voltage and load conditions. This is shown in Figure 2-12.

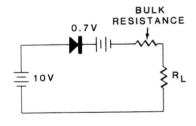

Figure 2-12 Circuit for Desk-Top Experiment 1

The simulated values of R_F are a theoretical concept to justify the sum of the voltage drops around a closed loop (Kirchoff's Voltage Law).

In truth, no current flows through the barrier junction until the 0.7 volts has been dropped. The 0.7 volts is the pressure required to force current to flow through the silicon junction. Once the 0.7 volts has been developed, any increase or decrease in current flow applies only to the load. Therefore when you see a PN junction in a circuit you should subtract its voltage drop and then analyze the remainder of your circuit.

22. The V-I characteristic curves of silicon and germanium diodes are similar.

    A. True
    B. False

23. A germanium diode will act as a low resistance in the forward direction when its forward voltage is greater than its internal _____ voltage.

24. A sudden increase in reverse current results when the _____ voltage of a diode is reached.

25. Reverse current is usually higher in _____ diodes.

26. The forward current in a diode is directly proportional (linear) to forward voltage when the forward voltage exceeds the diode's barrier voltage.

    A. True
    B. False

27. The abbreviation PIV stands for _____, _____, _____.

28. The forward voltage drop across a silicon diode will normally be higher than the forward voltage across a germanium diode when both diodes are conducting a relatively high forward current.

    A. True
    B. False

29. Germanium and silicon diodes cannot be damaged by excessively high reverse voltages since no current flows.

    A. True
    B. False

30. As the temperature of a diode decreases, the forward voltage drop:

    A. increases
    B. decreases
    C. remains constant

31. If a reverse-biased diode has a leakage current of 10 micro-amperes at a temperature of 25 degrees centigrade, what will its leakage current be if the temperature increases to 45 degrees centigrade?

    _____

32. A diode is forward-biased when the negative terminal of the battery is connected to the cathode of the diode and the positive terminal is connected the the anode of the diode.

    A. True
    B. False

33. The P section of a diode is represented by the _____ in the diode symbol.

Diode Construction

A PN junction diode cannot be formed by simply pressing a section of P-type material against a section of N-type material. It simply is not possible to obtain the intimate contact that is necessary between the two materials by using a method of this type. In actual practice, PN junctions are formed by using procedures that are somewhat complex although the basic principles involved are really quite simple. We will now briefly examine several of these basic construction techniques so that you will have a better understanding of how these junctions are actually created. Then we will see how these junctions are placed in suitable packages or containers to obtain a finished diode.

Grown Junctions

One of the earliest and most popular construction techniques is known as the grown method. With this method, PN junctions are actually grown by placing an intrinsic semiconductor and a P-type impurity into a quartz crucible (container) which is heated until the two materials melt. A small semiconductor crystal called a seed is then lowered into the molten mixture. The seed crystal is then slowly rotated and withdrawn from the molten mixture at a rate which will allow the molten material to cling to the seed. The molten material which clings to the seed crystal eventually cools and hardens and also assumes the same crystalline characteristics as the seed from which it appears to grow.

If the molten mixture contained only a P-type impurity as previously indicated, the grown crystal would be a P-type semiconductor. However, as the seed crystal is withdrawn, the molten mixture is alternately doped with N-type and P-type impurities. This means that N-type and P-type layers are created within the semiconductor crystal as it is grown. The resulting crystal is then sliced into many PN junctions.

Alloyed Junctions

Junctions may also be constructed by a technique known as the alloyed method. This method is extremely simple as it is performed by simply placing a small pellet of indium on the N-type semiconductor crystal as shown in Figure 2-13. The pellet and semiconductor are then heated until the pellet melts into and partially fuses with the semiconductor crystal as shown. Since indium is a trivalent impurity, the region where the two materials combine becomes a P-type semiconductor material as shown in Figure 2-13. Once the heat is removed the two materials recrystallize and a solid PN junction is formed.

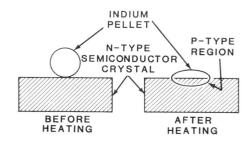

Figure 2-13 Alloyed PN Junction.

Diffused Junctions

One of the most popular and preferred techniques used to construct PN junctions is referred to as the diffusion method. With this method, a thin section of N-type or P-type semiconductor material sometimes called a wafer or die is exposed to an impurity element which is in a gaseous or vaporized state. The operation occurs at a very high temperature and the impurity atoms penetrate or diffuse through the exposed surfaces of the wafer. The basic diffusion process is shown in Figure 2-14. First a mask is placed over a N-type semiconductor crystal. Then the mask and the crystal are exposed to a gaseous trivalent impurity as shown. The mask has an opening which allows the gaseous impurity to strike a limited portion of the crystal's surface. The impurity element therefore diffuses into the N-type crystal and forms a P-type semiconductor material. The result is a PN junction as shown. The depth of the diffused impurity is controlled by regulating the time of exposure and the temperature of the materials. This diffusion

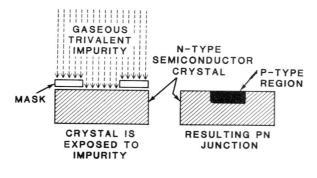

Figure 2-14 Diffused PN Junction.

process can also be performed by eliminating the mask and simply letting the entire semiconductor crystal be exposed to the gaseous impurity. However, it then becomes necessary to trim away the undesired portions of the crystal so that only the desired PN junction remains.

Diode Packaging

Once the PN junction is formed by using one of the three basic construction methods just described or by using similar or related techniques, the completed junction must be installed in a suitable container or package to produce the finished diode. This package protects the PN junctions from environmental and mechanical stresses and at the same time provides an efficient means of connecting the PN junction to other components or parts. The type of container or package selected for each PN junction is determined by the particular purpose or application of each diode. For example, if a PN junction is designed to handle large currents, it must be mounted in a package that will aid in this purpose. In other words the package must be designed so that it will keep the junction from overheating due to the high current flow. This type of PN junction diode is often constructed as shown in Figure 2-15. Notice that the semiconductor wafer (PN junction) is sandwiched between two metal discs and then attached to a heavy metal base which has a threaded bolt or stud. This stud serves as an electrical connection to one side of the junction, usually the N-side or cathode. A heavy metal lug provides an electrical connection to the other side of the junction. This lug is attached to the disc on the P-side or anode through an S-shaped wire. This flexible wire is bonded to the lug and to the disc and it insures that the PN junction is not subjected to the same stress and strain that might be placed on the metal lug. The metal lug is held in place by the glass top and is therefore, insulated from the metal base and the metal case. The diode is used by simply bolting it to a metal surface which will help to dissipate the heat from the device. Since the threaded stud is connected to the N-side of the junction, the stud serves as the cathode connection and any metal chassis, frame, or plate that comes in contact with the stud is electrically connected to the cathode. The metal lug connects to the P-side of the junction and therefore serves as the anode connection. A wire or component lead may be soldered to this lug to complete the installation of the diode.

The most common package for housing a PN junction diode is shown in Figure 2-16. The diodes contained in such packages are usually designed to handle small currents of 3 amperes or less. Most semiconductor diodes fall into this category. The semiconductor wafer (PN junction) is contained within a small cylindrical glass or epoxy case. The leads reach into the case to make contact with the wafer. The diode has axial leads and therefore can be installed in a variety of ways. In most cases the leads are simply soldered to a printed circuit board. The cathode end of the diode is usually identified by the band as shown. Sometimes a diode symbol is printed on the case to identify the leads.

The exact package and lead sizes depend on the type of diode, its application, and current handling capability. Most diodes are about 1/4″ long and 1/8″ in diameter.

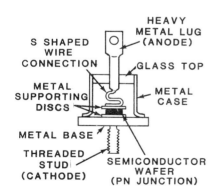

Figure 2-15 Cut-away view of a typical stud-mounted PN junction diode.

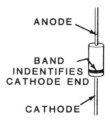

Figure 2-16 Typical semiconductor diode package.

34. A PN junction diode can be formed by tightly pressing N-type and P-type materials together.

    A. True
    B. False

35. The grown method of PN junction formation utilizes a _____ crystal which is dipped into a molten mixture of semiconductor materials and then withdrawn.

36. The construction technique which allows a pellet of idium to be fused to a semiconductor crystal is referred to as the _____ method.

37. The construction technique which allows a gaseous impurity element to penetrate a semiconductor wafer is referred to as the _____ method.

38. A _____ is sometimes used to control the location of the N-type or P-type region that is formed when a semiconductor wafer is subjected to a gaseous impurity.

39. A completed PN junction diode consists of a PN junction which is mounted in a suitable _____.

40. The cathode end of a diode may be identified by a _____ at one end of the device.

41. Diodes which use a heavy mounting stud or plate are usually designed to conduct relatively large forward currents.

    A. True
    B. False

Diode Applications

PN junction diodes are used extensively in the electronics industry. In some applications they are used simply because they are able to conduct current in the forward direction and block current in the reverse direction. However, in other applications they are used because of additional special characteristics that they possess. Many of the applications which utilize these special characteristics are discussed in the next two units. At this time we are primarily concerned with the diode's forward conduction and reverse blocking capabilities. This fundamental ability of the diode to pass current in only one direction is utilized in a process known as rectification. We will now briefly examine this rectification process which is often considered to be the most important application of diodes in electronics.

Rectification

Rectification is the process of converting alternating current (AC) into direct current (DC). Since diodes are used to accomplish this process they are often referred to as rectifiers. Rectification is a process that is required in most types of electronic equipment that must operate from alternating current. The primary power source for most electronic equipment is the AC power line. However, most electronic circuits usually require direct current for proper operation. The rectification process is therefore necessary to convert the available power (usually 110 or 220 volts, AC) into useable DC power which may be as low as several volts or as high as several thousand volts. The rectifier diode therefore does not operate alone in the rectification process. In fact a number of additional components are usually required. When assembled, these components form what is known as a power supply.

Figure 2-17 shows the basic rectification process. Notice that the input AC voltage is first applied to a transformer which either steps up the input voltage or reduces it to a lower value as required. The AC voltage at the secondary of the transformer is then applied to a diode and resistor which are in series. Although this resistor is shown as an individual component it actually represents the load which is the resistance of the electronic circuits which require DC power for operation.

Since the diode is in series with the load resistance, it will allow current to flow through the load in only one direction as shown. Current will flow through the diode and load resistance during each alternation of the input voltage that causes the diode to be forward biased. In other words, the diode will conduct on every other alternation of the input AC signal and therefore allow the current to flow through the load resistor in pulses. The pulsating load current (I_L) causes a pulsating output voltage to be developed across the load resistor as shown. This pulsating DC voltage is positive with respect to circuit ground because of the direction of the current through the resistor. The output voltage could be made negative with respect to ground by simply reversing the diode connections so that the direction of the current through the load resistor is reversed.

The pulsating DC output obtained from the rectifier circuit in Figure 2-17 does not usually provide an adequate operating voltage for most circuits. Most circuits require pure or continuous DC power for proper operation. This means that it is usually necessary to add extra components to the rectifier circuit to filter or smooth out the pulsations so that a constant output is obtained. The most commonly used filter component is a large capacitor connected across the load. The capacitor charges up to the peak value of the DC pulse during the time the diode conducts. During the next half cycle, the diode is reverse biased and does not conduct. During this time the capacitor discharges into the load to help maintain a continuous current flow. If the capacitor is large enough, the DC voltage across the load will be essentially constant.

The circuit shown in Figure 2-17 is known as a half wave rectifer since it conducts on only one half cycle of the AC input. Many other rectifier circuits are available to produce AC to DC conversion while providing other benefits as well.

Other Applications For Diodes

While rectification is by far the most common use of diodes, there are many electronic circuits that take advantage of the unilateral characteristics of the diode. For example, the rectifier circuit in Figure 2-17 is also used to demodulate amplitude modulated radio signals. This is the circuit that removes the audio signal from the carrier. Used in this way, the circuit is called a detector. Other diode detector circuits are used to demodulate FM and SSB signals and a variety of other signals.

Another application for diodes is waveshaping. This is the process of converting one type of signal into another. For example, a sine wave may be readily converted into a square wave by using diodes.

Diodes are also widely used for protection purposes. In this application diodes are used to prevent damage to circuits or other components.

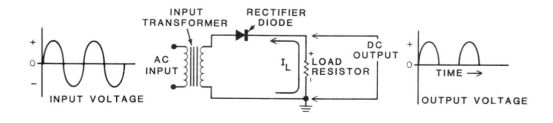

Figure 2-17 Typical rectifier circuit.

Diodes are often used as switches in automatic switching applications. Their polarity sensitive nature makes them easy to control in such situations.

Another application for diodes is in digital logic circuits. These are circuits that make decisions based on various input signals. Generally these circuits generate an output signal that is the result of a combination of input signals.

These are only a few examples of diode applications. In the next two units you will see further applications for special types of semiconductor diodes.

42. A diode used to convert AC into pulsating DC is called a:

_____.

43. The component most often used to filter the pulsating DC into pure DC is a:

_____.

44. A transformer, a rectifier diode and a filter are combined to form a circuit called a _____ _____.

45. Four other common applications for diodes are:

_____.

_____.

_____.

_____.

Unit Summary

Semiconductor diodes are produced by combining N-type and P-type semiconductor materials, forming a PN junction. The combination of these two materials allows free electrons and holes to combine until a depletion region, containing no majority carriers, is formed in the immediate vicinity of the PN junction.

The combination of electrons and holes causes positive charges to develop in the N-type material and negative charges that develop in the P-type material. These charges result from the ions that are left behind. Consequently, a barrier voltage develops across the junction. Eventually, the barrier voltage increases to a level that prevents further combining of electrons and holes, limiting the size of the depletion region. Germanium diodes have a typical barrier voltage of 0.3 volts. Similarly, a typical silicon diode barrier voltage is 0.7 volts.

When an external voltage that can overcome the barrier voltage is applied to the diode, the diode is forward biased, provided that the polarity of the voltage is correct. To forward bias a PN junction diode with a battery, the negative terminal must be connected to the N-type (cathode or banded) end of the diode and the positive battery terminal must be connected to the P-type (anode) end of the diode. A forward-biased diode conducts by means of majority carriers. The voltage drop across the PN junction of a forward-biased diode is equal to the barrier voltage.

Reversing the polarity of the voltage applied to the diode, so that the anode is negative with respect to the cathode and thus, reverse biases the diode. This is because the PN junction barrier voltage increases until it is equal and opposite to that of the applied voltage. This effectively cuts off almost all current through the diode. The small amount of reverse, or leakage, current through the diode consists of minority carriers and is usually insignificant. Conduction of a forward-biased diode is much heavier than that of a reverse-biased diode. Therefore, the forward resistance of a diode is much lower than its reverse resistance.

If the reverse bias voltage reaches a high enough value, the diode breaks down and conducts a relatively high reverse current. The voltage that causes this heavy reverse current is called the breakdown voltage of the diode. Heavy reverse current produces heat that can cause permanent damage to the diode unless a resistor is connected in series to limit reverse current to a level that is safe for the diode. The maximum reverse voltage that a diode can withstand without damage is called Peak Inverse Voltage (PIV). Diode manufacturers usually specify the PIV for each diode type.

Most diode characteristics are affected by temperature. Reverse leakage current and forward voltage drop are affected the most. In a typical diode, a ten-degree centigrade rise in temperature causes reverse current to nearly double. The forward voltage drop is inversely proportional to temperature; as temperature rises, the voltage drop decreases.

One method of forming a PN junction is by alternately doping intrinsic semiconductor material as a crystal is being grown from a seed. In another technique, an impurity is fused to material that is oppositely doped. However, the most common procedure used in the production of a PN junction is to inject an impurity into a previously doped semiconductor, forming a section of oppositely doped material.

After the PN junction is formed, electrical leads are added before the device is sealed in a protective package. The sealed package protects the PN junction from the environ-ment, while the leads provide the means to connect the diode to other components in a circuit. Several package styles are used for housing diodes, depending on the intended application. All of these packages provide a means of cathode and anode identification.

The diode's ability to conduct well in one direction while effectively blocking current in the other direction has many applications in electronic circuits. One application is as a rectifier for converting AC to DC. Other applications include detection, wave shaping, and logic and circuit protection.

Unit 3
The Zener Diode

Contents

Introduction

You are now ready to examine an electronic component that is closely related to the PN junction diode that you examined in the previous unit. This device is commonly referred to as a zener diode and it is widely used throughout the electronics industry. Because the zener diode is an extremely important electronic component, you must study this unit of instruction very carefully. By doing so, you will become familiar with another important solid-state component and at the same time expand your knowledge of semiconductor diodes in general.

Be sure to examine your unit objectives closely before reading the unit. Using this technique will improve your attention to detail and provide you with standards to evaluate your understanding as you proceed through this unit.

When you have completed this unit, you should be able to:

1. Explain the forward and reverse current-voltage characteristics of a typical zener diode.

2. Understand the relationship between temperature and zener diode power dissipation.

3. Know how to determine a zener diode's maximum safe operating current.

4. Know how to use a diode's zener impedance to determine the amount of change that can occur in the diode's zener voltage.

5. Describe how the zener diode is used to provide voltage regulation.

6. Identify a zener diode on a schematic diagram.

7. Explain the terms knee of the curve and reverse breakdown voltage.

8. Compare the zener diode to a junction diode in terms of internal voltage drop.

9. Draw the most common circuit configuration that uses a zener diode as a regulator.

10. Design a zener regulator circuit.

Zener Diode Characteristics

In the previous unit you learned that an ordinary PN junction diode breaks down and conducts a relatively high reverse current when it is subjected to a sufficiently high reverse bias voltage. This high reverse current occurs because the high reverse voltage is capable of tearing valence electrons away from their parent atoms and increasing the number of minority carriers in the N and P sections of the diode. The reverse voltage which causes an ordinary PN junction diode to shut down is the zener diode's operating voltage.

Ordinary PN junction diodes can be damaged if they are subjected to their respective breakdown voltages. This is because the high reverse currents produce more heat than the diodes safely dissipate. However, special diodes are constructed which can operate at voltages that equal or exceed their breakdown voltage ratings. These special diodes are called **zener diodes**.

We will now examine the exact relationship that exists between the current flowing through a zener diode and the voltage across the device. We will consider the action that takes place when the zener diode is forward-biased and reversed-biased, but we will be primarily concerned with the action that takes place at the point where breakdown occurs. Then we will see how a zener diode is rated according to its breakdown voltage.

Although very brief, this discussion on zener diode characteristics is very important. Pay particular attention to the new terms and symbols that are introduced. You will use these terms and symbols throughout this unit.

Voltage-Current Characteristics

A typical zener diode V-I(voltage-current) characteristic curve is shown in Figure 3-1. Notice that the overall forward and reverse characteristics of the zener diode are similar to those of an ordinary junction diode. The primary difference is that the zener diode is specifically designed to operate with a reverse bias voltage that is high enough to cause the device to breakdown and conduct a high reverse current. As shown in Figure 3-1, the zener diode's reverse current remains at a very low value until the reverse voltage is increased to a value that is sufficient to cause the diode to breakdown. Then the reverse current through the diode increases at an extremely rapid rate as the reverse voltage increases beyond the breakdown point. The V-I curve therefore shows that beyond the breakdown point, a very large change in reverse current is accompanied by only a very small change in reverse voltage. This action occurs because the resistance of the diode drops considerably as its reverse voltage is increased beyond the breakdown point. Once the breakdown point is reached the diode is operating in its zener region. The **zener current** may be represented by the symbol I_Z.

If you examine Figure 3-1 closely you will see that breakdown (also called zener breakdown) does not occur instantaneously. The curve is rounded near the breakdown point. This curved or rounded portion is often called the **knee** of the curve. When a zener diode has a knee with a very sharp edge, the diode enters the breakdown region very quickly. However, when the knee is more rounded, the breakdown region is entered at a slower rate. The importance of this consideration will be explained later in this discussion.

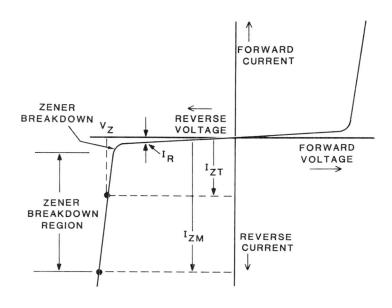

Figure 3-1 Typical V-I Characteristic Curve for a Zener Diode.

Zener Voltage

The breakdown voltage of a zener diode is determined by the resistivity of the diode which in turn can be controlled by the various doping techniques that are used to form the device. A zener diode is manufactured to have a specific breakdown voltage rating which is often referred to as the diode's **zener voltage** and is designated as V_Z. Typical V_Z values may vary from several volts to several hundred volts. For example, some of the popular low voltage units have ratings of 3.3, 4.7, 5.1, 5.6, 6.2, and 9.1 volts; however, additional voltage ratings are also available. It is important to realize that when a zener diode is rated as having a specific zener voltage (V_Z), the rated voltage does not represent the reverse voltage that is required to initially cause the diode to breakdown. The rated zener voltage is a nominal value that represents the reverse voltage across the diode when the zener current is at some specified value called the zener test current (I_{ZT}). The V-I curve in Figure 3-1 shows the relative zener voltage (V_Z) and **zener test current** (I_{ZT}) values for a typical zener diode. Notice that these values are located within the zener breakdown region of the curve. The zener test current (I_{ZT}) simply represents a typical value of reverse current that is always less than the maximum reverse current that the diode can safely handle.

Like resistors and capacitors, zener diodes cannot be produced with zener breakdown voltages that are always exactly equal to a specified value. Therefore, it is necessary to specify minimum and maximum breakdown voltage limits for each device. This is done by specifying a breakdown voltage tolerance for each type of diode that is manufactured. The standard zener breakdown voltage (zener voltage) tolerances are ± 20 percent, ± 10 percent, and ± 5 percent, however specially manufactured zener diodes are also available in 1 percent tolerance. For example, a 6.8 volt 10 percent zener diode will have a zener voltage somewhere in the 6.12 to 7.48 volt range.

Power Dissipation In Zener Diodes

Manufacturers of zener diodes also specify the maximum power dissipation of each device. Some devices are rated at only several hundred milliwatts while others are rated as high as 50 watts. However some of the most popular and widely used devices have relatively low ratings of 400 milliwatts, 500 milliwatts, and 1 watt. A zener diode's power dissipation rating is given for a specific operating temperature. Often the power rating is given for a temperature of 25 degrees centigrade, 50 degrees centigrade, or 75 degrees centigrade. However, the actual power that a zener diode can safely dissipate will decrease if the temperature increases above this specified level or increase if the temperature decreases below the specified level. Also, if the diode has axial leads, its power rating is often specified for a specific lead length or various ratings are given for various lead lengths. This is because a diode's ability to dissipate power increases as its leads are shortened. The shorter leads (when appropriately soldered in an electronic circuit) are more effective in conducting heat away from the diode's PN junction.

Power-Temperature Curves

To simplify the relationship between a zener diode's maximum power rating, its temperature, and its lead length, a power-temperature derating curve is often supplied with each type of diode manufactured. A typical curve for a diode that has a power dissipation rating of 500 milliwatts at a temperature of 70 degrees with lead lengths of 3/8″ is shown in Figure 3-2. Notice that three curves are shown for three different lead lengths of 1/8″, 3/8″, and 1″. The specified power rating of 500 milliwatts occurs only when the lead length is equal to 3/8″ and the temperature is equal to 75 degrees centigrade as shown. If the temperature increases above or decreases below 75 degrees centigrade, the power rating decreases below or increases above 500 milliwatts respectively. Also notice that the shorter lead length (1/8″) allows the diode to dissipate more power over the same temperature spread while the longer lead length (1″) reduces the overall power rating of the device. The curve also shows that the power rating of the device is effectively reduced to zero at 200 degrees centigrade. As you examine Figure 3-2 keep in mind that the actual temperature of the diode's leads are indicated and not just the ambient or surrounding air temperature which is sometimes shown in less specific power temperature curves. The diode's leads are also assumed to be soldered to a suitable circuit board or component which can serve as a heat sink to drain away the heat produced by the device.

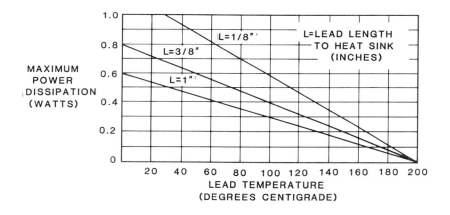

Figure 3-2 Power-Temperature Derating Curve.

3-8 SEMICONDUCTOR DEVICES

Derating Factor

If the zener diode is a high power type that is contained in a metal case or package that must be stud-mounted, a power-temperature derating curve may be provided which shows the diode's power rating for various case temperatures. However, in some cases the manufacturer will simply give a specific power rating for a zener diode (regardless of type) at a specific lead, case, or ambient temperature and then specify what is known as a **derating factor**. The derating factor is usually given in milliwatts per degree centigrade, and can be used to determine the power rating of the diode at temperatures that are different (usually higher) than the one specified. For example, a typical zener diode may have a derating factor of 6 milliwatts per degree centigrade. This simply means that the diode's power rating decreases 6 milliwatts for every degree centigrade increase in temperature.

1. Zener diodes are designed to safely operate within their zener breakdown regions.

   A. True
   B. False

2. The curved portion of the V-I curve near the point where zener breakdown occurs, is referred to as the _____ of the curve.

3. The current which flows through a zener diode that is operating in its zener breakdown region is referred to as the diode's _____ current.

4. Since it is impossible to manufacture zener diodes with breakdown voltages that are exactly equal to a specified value, it is necessary to indicate the minimum and maximum breakdown voltage limits by specifying a zener voltage _____ for each component.

5. A zener diode's power dissipation rating is usually given for a specific operating _____.

6. In general, the actual power that a zener diode can safely dissipate will decrease if temperature _____.

7. A diode's ability to dissipate power is increased when its leads are _____.

8. The relationship between a zener diode's maximum power rating, its temperature, and its lead length is often expressed graphically in the form of a _____ _____ _____ curve.

9. A zener diode's power rating for various temperatures can also be determined by using a _____ _____ which is usually given in milliwatts per degree centigrade.

Current Limitations In Zener Diodes

The maximum reverse current that can flow through a zener diode without exceeding the diode's power dissipation rating is commonly referred to as the **maximum zener current** and is represented in the V-I curve in Figure 3-1 by the symbol I_{ZM}. The I_{ZM} value of a zener diode is often specified by the manufacturer of the device. However, if I_{ZM} is not specified, it can be determined by simply dividing the power dissipation rating of the diode by its breakdown voltage (zener voltage) rating or stated mathematically:

$$I_{ZM} = \frac{\text{Power rating}}{\text{Zener voltage}}$$

However, it is best to play it safe and use the maximum limit of zener voltage in your calculations. For example, suppose you have a zener diode that is rated at 10 watts, and the diode has a zener voltage of 5.1 volts at ± 10 percent tolerance. The maximum voltage limit would be equal to 5.1 volts plus 10 percent of 5.1 volts or 5.1 + 0.51 which is equal to 5.61 volts. The maximum zener current would therefore be equal to:

$$I_{ZM} = \frac{10 \text{ watts}}{5.61 \text{ volts}}$$

$$= 1.78 \text{ Amperes}$$

The V-I curve in Figure 3-1 also shows that a small reverse or leakage current (I_R) flows through the zener diode before the break- down point is reached. Since the zener diode is normally used in its zener break- down region, this current is usually not im- portant. However, there are certain applica- tions of zener diodes which require an abso- lute minimum leakage current before the breakdown point is reached. Therefore man- ufacturers often specify the I_R value of a zener diode at a certain reverse voltage that is less than the zener voltage V_Z (often 80 percent of V_Z).

Effects Of Temperature On Zener Voltage

Zener diodes also have other characteristics that must be considered in certain applica- tions. For example, a diode's zener voltage will vary slightly as temperature changes. The amount of voltage change that takes place is usually expressed as a percentage of zener voltage (V_Z) change for each degree centigrade rise in temperature and is re- ferred to as the **zener voltage temperature coefficient**. Zener diodes that have a zener breakdown voltage of 5 volts or more, usu- ally have positive zener voltage temperature coefficients. This simply means that their breakdown voltages increase as temperature increases. However, most diodes that have breakdown voltages that are below approxi- mately 4 volts usually have a negative zener voltage temperature coefficient. This means that the breakdown voltage will decrease with an increase in temperature. When the breakdown voltages are between approxi- mately 4 and 5 volts, the zener voltage tem- peratures coefficient may be either positive or negative. For example, a zener diode with a zener breakdown voltage of 3.9 volts might have a zener voltage temperature coefficient of −.025 percent per degree centigrade. This means that the diode's zener voltage will decrease .025 percent (or approximately .001 volt) for each degree centigrade rise in temperature.

Temperature Compensated Zener Diodes

Special zener diodes are constructed which are temperature compensated so that their zener voltage ratings remain almost constant with changes in temperature. These special diodes are commonly referred to as temperature compensated zener diodes or voltage reference diodes. A temperature compensated diode is formed by connecting a zener in series with an ordinary PN junction diode. However, the two are connected back-to-back so that the junction diode usually has a zener voltage rating that is greater than 5 volts and therefore has a positive temperature coefficient. However, the forward-biased diode will have a forward voltage drop of approximately 0.6 or 0.7 volts and a negative temperature coefficient. By carefully selecting the two devices so that their temperature coefficients are equal and opposite, the voltage changes effectively cancel out. Furthermore, the voltage drops across the two devices must be summed to obtain the overall voltage rating of the temperature compensated device. For example, when a 5.6 volt zener diode is connected in series with a junction diode that has a forward voltage drop of 0.6 volts, a 6.2 volt temperature compensated zener diode is produced. In some cases more than one junction diode may be used to obtain the necessary compensation. Typical temperature compensated zener diodes will have temperature coefficients that range from .01 percent per degree centigrade to 0.0005 percent per degree centigrade. However, the optimum temperature stability usually occurs at or near a specific operating current which is normally specified by the manufacturer.

10. The maximum reverse current that can flow through a zener diode without exceeding the diode's power dissipation rating is referred to as the diode's _____ _____ _____.

11. The maximum safe value of reverse current that can flow through a zener diode can be calculated by dividing the diode's power rating by its _____ _____.

12. Manufacturers often specify the _____ current flow through a zener diode before its breakdown point is reached.

13. A diode's zener voltage temperature coefficient is usually expressed as a percentage of zener voltage change for each degree centigrade rise in _____.

14. A diode that has a zener breakdown voltage of 9.1 volts will have a _____ zener voltage temperature coefficient.

15. A diode that has a zener breakdown voltage of 3.3 volts will usually have a _____ zener voltage temperature coefficient.

16. When a diode's breakdown voltage decreases as temperature increases, the device has a _____ zener voltage temperature coefficient.

17. Temperature compensated zener diodes are formed by connecting a zener diode in series with one or more _____ _____ _____.

18. In a temperature compensated zener diode the zener diode portion of a device usually has a _____ temperature coefficient.

Zener Diode Impedance

Another important characteristic that should be considered when examining any type of zener diode is the diode's **zener impedance** (Z_{ZT}). This is determined by varying the zener current above and below the specified zener test current (I_{ZT}) value and then observing the corresponding change in zener voltage (V_Z) as shown in Figure 3-3. The zener impedance is equal to the change in zener voltage (V_Z) divided by the change in zener current (I_Z) and will vary considerably from one diode to the next.

Some zener diodes with low zener voltage ratings will have a Z_{ZT} of only a few ohms. In general, the lower the zener impedance, the greater the slope of the curve in the zener breakdown region. A low Z_{ZT} therefore indicates that the zener voltage changes only slightly with changes in zener current. An ideal zener diode would not change its breakdown voltage as zener current varies and would therefore have a zener impedance of zero ohms.

The zener impedance of a diode is also useful in determining the changes in zener voltage which can occur when the diode is used at zener currents which are higher or lower than I_{ZT}. It is a simple matter to calculate this change in zener voltage by using a known value of zener impedance and the change or deviation in zener current. When expressed mathematically, the change in zener voltage is equal to:

$$\Delta V_Z = \Delta I_Z \times Z_{ZT}$$

The same technique used to determine Z_{ZT} may also be used to determine the impedance at the knee of the curve near the point where breakdown occurs. The impedance at the knee of the curve is commonly referred to as the **zener knee impedance** (Z_{ZK}). The zener knee impedance provides an indication of the slope or sharpness of the knee of the curve. Manufacturers of zener diodes will usually specify both the zener impedance (Z_{ZT}) and the zener knee impedance (Z_{ZK}) for each device.

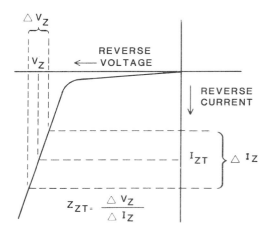

Figure 3-3 The Basic Method Used to Determine Zener Impedance (Z_{ZT}).

Zener Diode Packages

Basically, zener diodes are packaged in the same manner as the ordinary PN junction diodes described in the previous unit. The low power devices usually have axial leads and are mounted in either glass or epoxy cases while the high power units are usually stud mounted and are contained in metal cases. A typical low power zener diode is shown in Figure 3-4A and the commonly used zener diode symbol is shown in Figure 3-4B. A band is used to identify the cathode end of the zener diode as shown, therefore, the zener diode resembles an ordinary PN junction diode in appearance. The zener diode symbol is also similar to the ordinary diode symbol. The only difference is that the cathode end of the diode is represented by a zig-zag or Z-shaped bar instead of a straight bar.

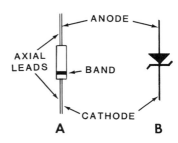

Figure 3-4 A Typical Zener Diode and its Schematic Symbol.

Voltage Regulation With Zener Diodes

Although the zener diode may be used to perform a number of important functions it is perhaps most widely used in applications where it is continually reversed-biased so that it operates constantly within its zener breakdown region. Under these conditions, the zener diode is effectively used to provide voltage stabilization or regulation.

Voltage regulation is often required because most solid-state circuits require a fixed or constant DC power supply voltage for proper operation. If this DC voltage changes significantly from the required value, improper operation will usually result. If an AC line operated power supply is used (one that is not regulated), the DC output voltage will vary if the power line voltage changes or if the load resistance connected to the power supply changes. However, by using a zener diode regulator circuit, it is possible to compensate for these changes and maintain a constant DC output voltage.

The Basic Zener Diode Regulator

A typical zener diode regulator circuit is shown in Figure 3-5. Notice that the diode is connected in series with a resistor and an unregulated (fluctuating or charging) DC input voltage is applied to these two components. The input voltage is connected so that the zener diode is reversed-biased as shown and the series resistor allows enough current to flow through the diode so that the device operates within its zener breakdown region. In order for this circuit to function properly, the input DC voltage must be higher than the zener breakdown voltage rating of the diode.

The voltage across the zener diode will then be equal to the diode's zener voltage rating and the voltage across the resistor will be equal to the difference between the diode's zener voltage and the input DC voltage.

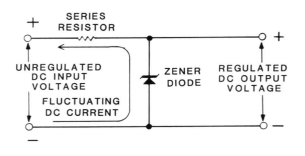

Figure 3-5 A Typical Zener Diode Voltage Regulator Circuit.

The input DC voltage shown in Figure 3-5 is unregulated or in other words it is not held to a constant DC value. This voltage may periodically increase above or decrease below its specified value and therefore cause the DC current flowing through the zener diode and the series resistor to fluctuate accordingly. However, the zener diode is operating within its zener voltage region and a wide range of zener currents can flow through the diode while its zener voltage changes only slightly. This tendency of the diode to oppose any change in voltage results because the resistance of the diode drops when zener current increases as described earlier. Since the diode's voltage remains almost constant as the input voltage varies, the change in input voltage almost completely appears across the series resistor which has a fixed or constant resistance.

You must remember that these two components are in series and the sum of their voltage drops must always be equal to the input voltage.

The voltage across the zener diode is used as the output voltage for the regulator circuit. The output voltage is therefore equal to the diode's zener voltage and since this voltage is held to a somewhat constant value, it is referred to as a regulated voltage. The output voltage of the regulator circuit can be changed by simply using a zener diode with a different zener voltage rating and selecting a series resistor that will allow the diode to operate within its zener breakdown region.

Voltage regulator circuits like the one shown in Figure 3-5 are used in many types of electronic equipment to provide constant operating voltages for various circuits. However, these circuits require not only operating voltage but various operating currents as well. Furthermore, there are many cases where operating currents continually vary because circuit impedances are not constant. When designing a regulator circuit it is therefore necessary to consider the specific current or the range of currents that the regulator must supply as well as the output voltage that the regulator must provide.

When a circuit, a lamp, or some other device is connected to the output of the regulator circuit in Figure 3-5, the regulator must supply current as well as voltage to the external device (commonly called a load). This situation is shown in Figure 3-6. Notice that the load resistor (R_L) requires a specific load current (I_L) which is determined by its resistance and the output voltage. The current through the zener diode (I_Z) combines with I_L and flows through the series resistor (R_S). The value of R_S must therefore be chosen so the I_Z remains at a sufficient level to keep the diode within its breakdown region and at the same time allow the required value of I_L to flow through the load.

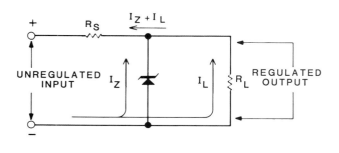

Figure 3-6 Loading Affects in a Zener Diode Voltage Regulator.

If the load current increases or decreases because R_L effectively decreases or increases in value, it might appear that the output voltage would change accordingly. In other words even if the input voltage remains constant there is a tendency for the output voltage to increase or decrease if R_L increases or decreases in value respectively. If R_L and R_S were in series and the zener diode was not installed in the circuit, this action would definitely result. However, the zener diode prevents this from happening. When R_L increases in value, I_L decreases and the voltage across R_L tries to increase. However, the zener diode opposes this change by conducting more zener current (I_Z) so that the total current ($I_Z + I_L$) flowing through R_S

remains essentially constant (assuming that the input voltage remains constant). The same voltage is therefore maintained across R_S and also across the parallel branch (the zener diode and R_L). Also, whenever I_L increases, I_Z decreases by approximately the same amount to again hold $I_Z + I_L$ almost constant and therefore maintain an essentially constant output voltage. In this way the zener diode regulator is able to maintain a relatively constant output voltage even though changes in output current occur. The circuit therefore regulates for changes in output current as well as for changes in input voltage.

Designing a Zener Regulator Circuit

To further illustrate the use of the zener diode in a typical voltage regulator circuit we will now design a typical regulator using some basic design rules. We will begin by assuming that certain input and output conditions must exist and we will calculate the necessary component values. First we will assume that the unregulated input voltage varies from a minimum value of 9 volts to a maximum value of 12 volts. The regulated output voltage will be equal to 5.1 volts which is a standard zener voltage rating. The output load current will vary from 0 to 30 milliamperes. To satisfy these conditions we will of course need a zener diode with a zener voltage rating of 5.1 volts; however, we must also use a series resistor (R_S) that will allow the diode to operate properly with the above mentioned changes in input voltage and output current. The required value of R_S can be quickly determined by using the following equation.

$$R_S = \frac{V_{in(min)} - V_Z}{1.1\ I_{L(max)}}$$

This equation states that R_S is equal to $V_{in(min)}$ (the minimum value of input voltage) minus V_Z (the zener voltage) divided by 1.1 $I_{L(max)}$ (1.1 times the maximum value of load current). By subtracting V_Z from $V_{in(min)}$ we obtain the minimum voltage that will be dropped across R_S. This voltage must then be divided by 1.1 $I_{L(max)}$ to determine the required value of R_S. Multiplying 1.1 times $I_{L(max)}$ is the same as increasing the value of $I_{L(max)}$ by 10 percent. This additional current is used as a safety factor to insure that the zener diode's current does not drop below the level needed to keep the diode operating within its zener breakdown region when the unregulated input voltage is at its minimum value. You must remember that the total current is equal to I_Z + I_L and when I_L is at a maximum value, I_Z must decrease to a minimum value. By specifying a maximum load current that is 10 percent higher than required, we are insuring that this additional current (which is not used by the load) will always flow through the diode.

When we substitute actual values in the equation shown above we find that the series resistance must be equal to:

$$R_S = \frac{9\ V - 5.1\ V}{1.1\ (0.3)\ A}$$

$$= 118\ \Omega$$

Since this is not a readily available standard resistance value, we must select the next lower value of resistance available which is 100 ohms. A lower value is selected to insure that I_Z will not drop below the level necessary to keep the diode operating within its zener breakdown region.

Now we must determine the maximum power that the zener diode will be required to dissipate. The maximum power dissipated by the zener diode can be easily calculated by using the following equation.

$$P_{Z(max)} = V_Z \left[\frac{V_{in(max)} - V_Z}{R_S} - I_{L(min)} \right]$$

This equation simply states that $P_{Z(max)}$ (the maximum power dissipated by the zener diode) is equal to V_Z (the zener voltage) times the maximum value of zener current. However, the maximum zener current is determined by finding the maximum current through R_S and then subtracting $I_{L(min)}$ (the minimum load current) from this maximum current. The maximum current through R_S is found by subtracting V_Z (the zener voltage) from $V_{in(max)}$ (the maximum input voltage) and dividing the difference by R_S.

When we substitute actual values into this equation we find that the zener diode must dissipate a maximum power of:

$$P_{Z(max)} = 5.1 \text{ V} \left[\frac{12 \text{ V} - 5.1 \text{ V}}{100 \text{ }\Omega} - 0 \right]$$

$$= 0.352 \text{ watts}$$

In this circuit you would use a zener diode with a power dissipation rating that is slightly higher than 0.352 watts (352 milliwatts) to provide an additional safety margin. If the diode is to be used where the ambient temperature is near 25 degrees centigrade (normal room temperature) we could select a diode that is rated at either 400 milliwatts or 500 milliwatts at this temperature.

Throughout all of our calculations we have assumed that the zener voltage V_Z remains essentially constant. This simplifies our calculations and in most cases it is a practical assumption. In cases where the actual changes in V_Z (usually only several tenths of a volt) must be considered, the diode's zener impedance (specified by the manufacturer) is helpful in determining these changes as zener current varies. It is also important to select a diode with a zener voltage tolerance that is suitable for the particular application. A ±20 percent tolerance might be acceptable in a non critical application, but a lower tolerance of ±5 percent or ±1 percent may be needed where the output voltage must be very close to the specified value.

Desk-Top Experiment 2
Designing a Zener Diode Regulator

Introduction

This experiment is designed to ensure that you understand the operation of the zener diode when it is used as a voltage regulator. It will also demonstrate the zener diode's relationship to the other circuit components.

Objective

Design a zener diode shunt regulator that has an output of -6.2 volts at approximately 60 mA of load current from a voltage source that varies from -8 to -12 volts.

Procedure

1. Draw the zener diode shunt regulator circuit described in the objective.

2. Compute the required value of R_S to meet the specifications stated in the objective.

3. Compute the required value of R_L to meet the specifications stated in the objective.

4. Compute $P_{Z(max)}$.

5. Calculate the voltage drops across R_S for input voltages of 8, 9, 10, 11, and 12 volts. Write your answers in the table provided in Figure 3-7.

6. Calculate the values of I_Z for 8, 9, 10, 11, and 12 volts. Write your answers in the table provided in Figure 3-7.

	VOLTS				
V_{in}	8	9	10	11	12
E_{R_S}			3.8		
E_{R_L}			6.2		
I_Z			75		
$P_{Z(MAX)}$			465		

Figure 3-7 Results of Desk-Top Experiment 2 Steps 1—4

Discussion

The shunt regulator circuit that you drew should have looked like the circuit shown in Figure 3-8. The values of R_S, R_L, and $P_{Z(max)}$ can be determined using the equations:

$$R_S = \frac{V_{in(min)} - V_Z}{1.1 \, I_{L(max)}}$$

$$R_L = \frac{V_Z}{I_L}$$

$$P_{Z(max)} = V_Z \left[\frac{V_{in(max)} - V_Z}{R_S} - I_{L(min)} \right]$$

The values of the voltage drops across R_S and R_L for an input voltage of 10 volts have been included in Figure 3-7. This is to ensure that your calculations are correct and that you are using the proper procedures for designing a shunt regulator.

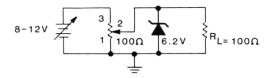

Figure 3-8 Circuit for Desk-Top Experiment 2 Summary

Procedure (Cont.)

At this point you may think that you have completed the experiment, but there are a few more considerations before the circuit design is complete.

7. You have determined the resistance values of R_S and R_L. Have you determined their physical size requirements?

8. Calculate the power dissipation requirements for R_S and R_L.

   I^2R value:

   R_S dissipates _____.

   R_L dissipates _____.

9. List all of the component parts required to build your circuit and include their specifications.

   A.
   B.
   C.

10. Assume that the zener diode is reversed in the circuit and no other changes are made. What would the output voltage equal?

   $$E_{(out)} = \text{_____}.$$

Discussion

Knowing the value of the resistors required in a circuit design is only half the battle. The other consideration is circuit protection and that means, that all of the components must be able to handle the currents and voltages that they could be exposed to. It is a good practice to use the next higher standard size component instead of the exact calculated value. This provides more component protection, and by using standard sizes you will reduce the cost of your circuit. When you use standard value components, it increases the number of sources where replacement parts can be obtained.

If the zener diode is physically reversed in the circuit it would be forward biased. The output voltage would be the standard 0.7 volt that is always dropped across a conducting PN junction.

Summary

Your first consideration is that the source voltage must be negative and capable of providing an input voltage of more than minus 8 volts, but not more than minus 12 volts.

Next R_S must be capable of handling the maximum possible current. That means that R_S must be able to handle the currents flowing through both the zener diode and the load resistor. This is because R_S is in series with both components.

R_L must be able to dissipate the heat generated by the load current and the zener diode must be able to handle the maximum circuit current just in case the load opens.

$$\text{Maximum current} = I_Z + I_L$$

If the zener diode is reversed it would also be necessary to reverse the polarity of the source voltage to convert your negative voltage regulator into a positive voltage regulator. The values and physical sizes of R_S and R_L are valid for a positive regulator.

19. A diode's zener impedance is determined by dividing a change in _____ _____ by a corresponding change in zener current.

20. An ideal zener diode would have a zener impedance of _____ ohms.

21. The impedance at the knee of a zener diode's V-I curve is referred to as the diode's _____ _____ _____ .

22. Low power zener diodes usually have _____ leads.

23. High power zener diodes are usually _____ mounted.

24. The zener diode symbol is similar to the PN junction diode symbol but has a _____ shaped bar instead of a straight bar.

25. A zener diode is often used in applications where it is continually _____-biased.

26. When used in a voltage regulator circuit the zener diode must be subjected to an input DC voltage that is higher than the zener breakdown voltage rating of the diode.

   A. True
   B. False

27. The output voltage produced by the voltage regulator circuit in Figure 3-5 is approximately the zener diode's _____ _____ rating.

28. The voltage across the series resistor in Figure 3-5 remains constant as the unregulated input voltage varies.

   A. True
   B. False

29. The circuit shown in Figure 3-6 provides a relatively constant output voltage even though changes in input voltage and changes in _____ _____ occur.

30. The zener current (I_Z) in Figure 3-6 should _____ when the load current (I_L) increases so that the output voltage will remain constant.

31. Design a zener regulator whose output voltage is 12 volts. The load current is fixed at 40 mA but the input voltage varies over a 15 to 20 volt range. Calculate R_S and the power dissipation of the zener.

PN junction diodes that are intended for use as rectifiers can be damaged if they are subjected to reverse voltages that equal or exceed their breakdown voltage ratings. This damage results from heat produced by the heavy reverse current through the device.

Zener diodes are designed to withstand the high reverse current that results when the diode's breakdown voltage is applied. After the breakdown voltage of the zener diode is exceeded, reverse current through the diode increases from an insignificant leakage value to a relatively high value. Once a zener diode is operating in its breakdown, or zener region, an increase in current causes the internal resistance of the diode to decrease. Conversely, if current decreases, the diode's internal resistance increases. As a result, the voltage drop across the zener diode remains essentially constant. Current through a zener diode operating in the zener region is called zener current. If zener current exceeds the specified maximum for a given diode, the diode can be permanently damaged.

You can determine the zener voltage rating (V_Z) of a given zener diode by measuring the voltage drop across the diode with a specified zener test current (I_{ZT}) through the diode. Zener test current is always greater than the reverse current that initiates zener breakdown, but less than the maximum current that the device can withstand.

Zener diodes are available for almost any zener voltage rating. However, it is impractical to manufacture a zener diode capable of dropping an exact specified voltage. Therefore, a certain amount of variation is expected. Consequently, manufacturers usually specify the amount of variation expected as a percentage of tolerance. For example, if a given zener diode is rated at 9.1 volts with a tolerance of $\pm 20\%$, the actual voltage dropped by that diode may be any voltage in the range of approximately 7.3 volts to 10.9 volts.

Zener diode manufacturers also specify the maximum power that their products can safely dissipate at a given temperature. Diode power ratings increase or decrease with decreases or increases in temperature, respectively. In addition, you can increase the power rating of a zener diode with axial leads by shortening the length of the leads.

The most common application of zener diodes is as regulators in regulated DC electronic power supplies. When a zener diode is used as part of a properly designed regulator circuit, the diode maintains a constant (regulated) DC output voltage. In a typical regulator circuit, the diode is connected in series with a resistor to the unregulated output of the power supply. The load is then connected in parallel with the diode. Consequently, any variations in the unregulated output voltage caused by load current variations or line voltage changes are dropped across the resistor, while the voltage across the zener diode and the load remains constant. In addition, the sum of the zener and load currents remains fairly constant.

Unit 4
Semiconductor Diodes For Special Applications

Contents

Introduction

Now that you have examined the basic PN junction diode and the zener diode, it is time to examine some other types of diodes that have unique characteristics that make them suitable for special applications. Some of these special diodes are constructed in the same basic manner as the PN junction diodes previously described while others are formed by using entirely different construction techniques. Study this unit carefully as it will greatly expand your knowledge of semiconductor diodes and emphasize the important role that these devices play in electronics.

Examine your unit objectives carefully as they are the key to the material that you are expected to learn and retain in this unit.

When you have completed this unit on special semiconductor diodes you should be able to:

1. Locate the peak current, valley current, peak voltage, and valley voltage points on a tunnel diode's V-I curve.

2. Explain the term negative resistance.

3. Identify the negative resistance region of a tunnel diode's characteristic curve.

4. Identify commonly used tunnel diode schematic symbols.

5. Explain how capacitance is varied within a varactor diode.

6. Understand how a varactor's internal capacitance and Q are affected by a change in operating voltage.

7. Identify commonly used varactor diode symbols.

8. Name four special diode types other than tunnel and varactor diodes, and state an application for each.

9. From a schematic diagram identify a:

   PN junction diode
   Zener diode
   Tunnel diode
   Varactor diode
   Schottky diode

The Tunnel Diode

The ordinary PN junction diodes and zener diodes that you already examined have lightly doped PN junctions and they have voltage-current characteristics that are quite similar. However, there is a special type of PN junction diode that is produced by using special construction techniques and heavy doping (a high concentration of impurities) to obtain characteristics that are radically different from those of ordinary PN junction or zener diodes. This device is commonly referred to as a **tunnel diode**.

Due to its heavily doped PN junction, the tunnel diode has a high internal barrier voltage and an extremely narrow depletion region. The device also has an extremely low reverse breakdown voltage (almost zero) and therefore conducts large currents when it is reversed-biased. The forward characteristics of the device are also unique. It would appear that the high barrier voltage would prevent forward current from flowing through the diode when it is subjected to low forward bias voltages; however, this is not the case. At low forward bias voltages, electrons are forced through the narrow depletion region at an extremely high velocity because of the high concentration of charges on each side of the junction. The electrons effectively appear to tunnel through the potential barrier (barrier voltage) as they move across the junction. Also, during this period of time when the tunneling action is occurring, a point is reached where an increase in forward voltage can actually cause a decrease in forward current through the diode. This movement of electrons can be explained by a theory known as quantum mechanical tunneling; however, a detailed account of this theory is not necessary at this time. It is sufficient to say that when voltage increases and current decreases is contrary to Ohm's Law. Thus, the term negative resistance is used to describe this phenomenon.

Voltage-Current Characteristics

A typical tunnel diode V-I characteristic curve is shown in Figure 4-1. As shown in this figure, the diode will conduct high reverse currents when it is subjected to reverse voltages. However, the diode's most important electrical characteristics occur when it is forward-biased. Notice that the forward current through the diode initially increases as the forward bias voltage is increased, but a point is soon reached where forward current stops increasing with an increase in forward voltage. The current that flows through the tunnel diode at this time is referred to as the **peak current** I_p and the voltage across the diode is referred to as the **peak voltage** V_p. A further increase in forward voltage causes the forward current through the diode to decrease as shown. The current continues to decrease as forward voltage is increased until it reaches a minimum value which is referred to as the **valley current** I_v. At this time the voltage across the diode is referred to as the **valley voltage** V_v. If the forward voltage across the diode is increased still further, the diode's forward current will again increase. However, this time the current increases in the same way as it would in an ordinary PN junction diode that is subjected to an increasing forward bias voltage.

Between the peak and valley points on the V-I curve, the tunnel diode's forward current decreases as the forward voltage across the device is increased. This portion of the diode's V-I curve is therefore referred to as the **negative resistance** region. The tunneling action previously mentioned is reduced throughout this portion of the curve and it ceases when the I_V value is reached. It is this negative resistance region that makes the tunnel diode an extremely useful electronic component.

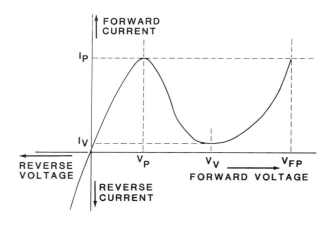

Figure 4-1 Typical tunnel diode V-I Characteristics.

What is Negative Resistance?

Negative resistance is that characteristic of an electronic component or circuit where the current through it decreases as the voltage across it increases and vice versa. The resistance of the component or circuit is not really negative in the true sense of the word. Its resistance is a positive value, but its effect defies Ohm's law as we know it. Ohm's law says that increasing the voltage across a resistance will result in an increase in the current through that resistance.

$$E = IR$$

Manufacturers of tunnel diodes usually specify the values of peak current I_P, peak voltage V_P, valley current I_V, and valley voltage V_V for each device. These values must be known in order to properly bias the tunnel diode within its negative resistance region. The peak current can be easily regulated regardless of the type of semiconductor material used. Most tunnel diodes are designed to have low peak currents (often as low as 100 microamperes) but devices have been constructed that have peak currents as high as 10 amperes. The valley current is usually held to a low value with respect to the peak current so that a high peak-to-valley current ratio I_P/I_V is maintained. The greater this ratio, the greater the operating current range within the negative resistance region. The peak voltage and valley voltage are determined by the type of semiconductor material used to construct the diode and for all practical purposes have fixed values. For example, a germanium tunnel diode will have typical V_V and V_P values of 55 and 350 millivolts at 25 degrees centigrade.

In addition to the V_P and V_V values, the manufacturer may also specify the **projected peak value** V_{FP}. This is the forward voltage at which the forward current rises to a point where it is again equal to I_P as indicated in Figure 4-1. The V_{FP} value of a particular diode is also determined by the type of material used in its construction. For example, germanium devices will have a typical V_{FP} of 500 millivolts at 25 degrees centigrade.

Manufacturers sometimes define the negative resistance region of a device as a **negative conductance** $-Gb$. The negative conductance is determined by dividing a change in forward current (delta I) by a corresponding change in forward voltage (delta V) and is expressed in a unit known as the mho. The negative conductance of a device therefore provides an indication of the slope (rate of change of current with respect to voltage) of the curve within the negative resistance region. In some cases the manufacturer will specify the reciprocal of the negative conductance or the **negative resistance** $-R_d$ of the device. The negative resistance of the diode is therefore equal to $1/-G_d$ and is expressed in ohms.

For example, a tunnel diode may be manufactured to have a negative conductance that is within minimum and maximum limits of 0.0065 to .01 mho. This same diode would therefore have a negative resistance that varies from 100 to 150 ohms.

Construction

Tunnel diodes may be constructed from sev-

eral types of semiconductor materials. Basic semiconductor materials such as germanium and silicon may have been used for years but many of the newer devices are constructed from gallium arsenide and gallium antimonide. The tunnel diode PN junctions may be formed by using the grown method, the diffusion method, or the alloyed method. However, the alloyed method is perhaps the most widely used construction technique. Many tunnel diodes resemble ordinary PN junction diodes or zener diodes in appearance but some tunnel diodes are packaged in special cases or containers which make them suitable for various applications. A typical tunnel diode and its approximate dimensions are shown in Figure 4-2. This particular diode is packaged in a metal case and has a peak current rating in the low milliampere range. Since the device is only one tenth of an inch long and one tenth of an inch wide, it is smaller than the head of a match.

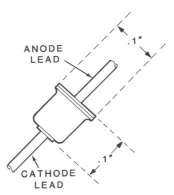

Figure 4-2 Typical tunnel diode package.

Several commonly used tunnel diode symbols are shown in Figure 4-3. Two of the symbols resemble the conventional PN junction diode symbol, but the third symbol is completely different. This symbol consists of a bar; which represents the anode, and a half circle, which represents the cathode. This symbol is also occasionally drawn within a full circle like the diode symbol shown on the left.

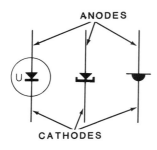

Figure 4-3 Tunnel diode symbols.

Applications

The tunnel diode is particularly suitable for use in oscillator circuits which are designed to generate high frequency AC signals. A typical tunnel diode oscillator circuit is shown in Figure 4-4. When used in this manner the diode must be biased so that it will operate within its negative resistance region.

Notice that the diode is connected in series with an LC resonant circuit. The circuit receives its power from the battery, and the

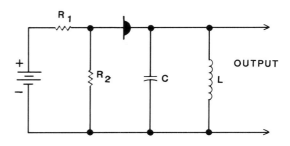

Figure 4-4 Typical tunnel diode oscillator circuit.

two resistors (R_1 AND R_2) are used to set the diode's operating current and voltage into the negative resistance region.

The resonant LC circuit cannot sustain oscillations when it is used alone. However, when the LC circuit is used with the tunnel diode as shown, continuous oscillations are produced and an AC output voltage can be taken from the resonant circuit. The continuous oscillations result because of the diode's negative resistance. When power is first applied to the circuit, oscillations are produced within the LC circuit. These oscillations produce a voltage across the LC circuit and this voltage alternately causes a shift in the diode's operating point. This in turn causes the diode's resistance to change in a manner which allows the current through the diode to reinforce the circulating current that flows through the resonant LC circuit.

The tunnel diode's negative resistance is used to support the oscillations that are produced within the LC circuit and the power losses within the LC circuit are effectively reduced to zero.

The tunnel diode may also be used as an electronic switch. When used in this manner the device is made to change between two-operating states. In one state the device conducts a relatively high forward current at a point just before I_P is reached. In the other state it conducts a relatively low current at a point just beyond the I_V value of the device. Tunnel diodes have been used to implement switching circuits that perform high speed digital logic functions.

When used in either of the applications just described the tunnel diode requires very little power and it is capable of operating at very high speeds. When used as an oscillator the tunnel diode is capable of operating in the microwave frequency range (above 200 megahertz) and when used as a switch the device can change states in only a few nanoseconds (10^{-9}). Unfortunately, tunnel diodes also have disadvantages which have seriously limited their use in many applications. In general, the tunnel diode's important electrical characteristics vary widely with changes in temperature and their operation is greatly affected by changes in operating voltages. These two factors make it difficult to stabilize the operation of tunnel diode circuits.

1. A tunnel diode will conduct high reverse currents when it is reversed-biased.

   A. True
   B. False

2. A tunnel diode's negative resistance region occurs when the device is _____-biased.

3. Throughout the tunnel diode's negative resistance region, forward current _____ as forward voltage increases.

4. A tunnel diode's peak current Ip represents the maximum current that the diode can safely handle.

   A. True
   B. False

5. The minimum forward current that flows through a tunnel diode before it functions like a conventional diode is referred to as the diode's _____ current.

6. Tunnel diodes are usually designed to have a high _____ to _____ current ratio.

7. The negative conductance of a tunnel diode provides an indication of the _____ of the diode's V-I curve within the negative resistance region.

8. A tunnel diode's projected peak voltage (V_{FP}) is the forward voltage that appears across the diode when the diode's forward current rises to a value that is again equal to the diode's _____ current.

9. A tunnel diode's negative resistance can be determined by taking the reciprocal of its negative conductance.

   A. True
   B. False

10. When used in an oscillator circuit, the tunnel diode is biased to operate within its _____ _____ region.

The Varactor Diode

When any PN junction diode is reverse-biased, majority carriers are swept away from the diode's junction and a relatively wide depletion area is formed. When the diode is subjected to varying reverse bias voltage, the width or thickness of this depletion region will also vary. When the reverse bias voltage increases in value, the depletion region becomes wider. When the reverse voltage decreases, the depletion region becomes narrower. The depletion region acts like an insulator since it provides an area through which no conduction can take place. It also effectively separates the N and P sections of the diode in the same way that a dielectric separates the two plates of a capacitor. In fact, the entire PN junction diode is basically a small electronic capacitor that changes its capacitance as its depletion region changes in size. Remember that when the plates of a capacitor are moved further apart, capacitance decreases.

Ordinary PN junction diodes possess only a small amount of internal junction capacitance and in most cases this capacitance is too small to be effectively used. However, special diodes are constructed so that they have an appreciable amount of internal capacitance and are used in much the same way that ordinary capacitors are used in electronic circuits. These special diodes are commonly referred to as varactor diodes or simply varactors.

Electrical Characteristics

A varactor is usually operated with a reverse bias voltage that is less than its reverse (zener) breakdown voltage rating. As the reverse voltage is increased, the depletion region within the device widens and therefore acts as a wider dielectric between the N and P sections of the device. Since the value of any capacitor varies inversely with the thickness of the dielectric between its plates, the diode's junction capacitance will decrease as the reverse voltage increases. This means that a decrease in reverse bias voltage will cause an increase in the varactor diode's internal junction capacitance.

The varactor diode's capacitance therefore varies inversely with the reverse bias voltage applied to the device. However, the capacitance does not vary in a linear manner and therefore is not inversely proportional to the reverse bias voltage. The nonlinear change in capacitance that occurs in a typical varactor diode is shown in Figure 4-5. Notice that capacitance drops rapidly from a maximum value of approximately 40 picofarads as the reverse bias voltage is increased from zero volts. However, the capacitance levels off to a value of approximately 5 picofarads when the reverse bias voltage approaches 60 volts. Although not shown in Figure 4-5, the capacitance of the diode will actually increase above 40 picofarads if the diode is forward biased and will continue to increase as forward bias is increased. However, a point is soon reached where the diode's barrier voltage is overcome (0.6 or 0.7 volts for silicon diodes) and the diode will conduct a high forward current.

This forward current makes the diode useless as a capacitor and operation in the forward conduction region must be avoided. Also, should the reverse breakdown voltage of the diode be reached the device would again become useless as a capacitor because of the high reverse current that would occur.

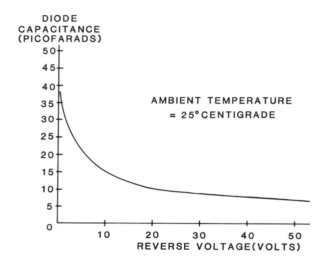

Figure 4-5 Typical capacitance versus voltage characteristics of a varactor diode.

A varactor's internal capacitance will increase or decrease slightly when temperature increases or decreases. Therefore, manufacturers of these devices will usually show a range of capacitance values at a particular operating temperature. For example, the curve shown in Figure 4-5 was plotted at an ambient temperature of 25 degrees centigrade. The amount of capacitance change that can be expected for a given change in temperature is usually expressed

as a temperature coefficient of capacitance TC_c. The TC_c is expressed in percent of capacitance change per degree centigrade.

Operating Efficiency

The relative efficiency of a capacitor is expressed as a ratio of the energy stored by the device to the energy actually used or dissipated by the device. This ratio is referred to as the quality factor or Q of the capacitor and it can be determined by dividing the capacitive reactance X_C of the capacitor by the series resistance R_s of the device. However, X_C varies with frequency and the Q will therefore change as frequency changes. Since $X_C = 1/2\pi fC$, the Q can be expressed in more specific mathematical terms as:

$$Q = \frac{1}{2\pi fCR_s}$$

A varactor diode has internal capacitance and as far as an AC signal is concerned the device also has an internal resistance. This internal resistance is due to the bulk resistivity of the semiconductor material and even though the device is reverse-biased and effectively blocks DC current, this internal resistance is seen by AC signals which pass through the device because of its capacitance. The internal resistance appears as a small resistance in series with the device. This means that varactors have a measurable Q just like ordinary capacitors. In fact, most varactors have very high Q's.

The equation for Q shows that the Q of any capacitor decreases as frequency, capacitance, or series resistance increases. Therefore, the Q of a varactor must be specified for particular operating conditions if it is to provide a meaningful indication of the diode's efficiency. Since the capacitance of a varactor varies inversely with reverse voltage, the Q of the device will increase as the reverse bias voltage is increased. The maximum Q is therefore obtained just before the breakdown voltage is reached. Often, manufacturers will provide graphs which plot Q values as well as capacitance values over a range of reverse bias voltages. In other cases manufacturers will simply indicate the Q of a particular device at a specific frequency (often 50 megahertz) and a specific reverse bias voltage. When comparing Q values of various varactor diodes, always remember that the specific operating conditions must also be considered.

If a varactor is subjected to a sufficiently high frequency the Q of the device can be reduced to a value of 1. The frequency at which this point occurs is commonly referred to as the cutoff frequency f_{co} of the device. This cutoff frequency may be determined for various values of C and R_S by using the following equation:

$$f_{co} = \frac{1}{2\pi CR_s}$$

Manufacturers of varactor diodes will often specify the maximum leakage current I_R of each device since current could have a serious effect on the Q of the device and the manner in which it will operate in various circuits. This leakage current is usually extremely small (in the low micrampere range). In order to obtain a high Q, the leakage current must be very low. For this reason most varactor diodes are made from silicon although some units that are designed to operate at microwave frequencies are made from gallium arsenide.

Construction

Varactor diodes are designed for a number of specific applications. They come in a variety of sizes and shapes and have various power, current, and voltage ratings. These diodes may have capacitance values that range from less than 1 picofarad to one or two thousand picofarads. General purpose, low power diodes are available which closely resemble the ordinary PN junction diodes described in a previous unit. These low power units are often rated at several hundred milliwatts and they are usually packaged in a glass case with axial leads. Other types are available that have higher power ratings (25 to 35 watts) and are stud mounted just like power diodes. These are commonly used at frequencies below 500 megahertz (MHz). At frequencies much above 500 megahertz the physical layout of the diode package becomes extremely important. Therefore most of the diodes that operate at these higher microwave frequencies are usually contained in packages that are designed to minimize stray or unwanted capacitance and inductance which can cause undesirable effects at these frequencies. Two varactor diodes that are designed to operate at frequencies well above 500 megahertz are shown in Figure 4-6. These devices are extremely small, as indicated by the approximate dimensions shown, and have power dissipation ratings of 500 milliwatts.

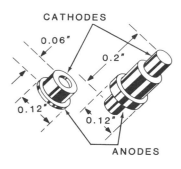

Figure 4-6 Typical high frequency varactor diode package.

It is important to realize that the power dissipation rating of a varactor diode does not have the same meaning as it does in the case of an ordinary PN junction diode. The power dissipated by an ordinary diode is equal to the product of the diode's forward DC current and its forward voltage. However, a reverse-biased varactor diode effectively does not allow a reverse current to flow, but it will pass an AC current. It is this AC current that determines the amount of power dissipated by the device because it flows through the series resistance (R_s) of the device and causes heat to be generated. The power dissipated by the device is therefore based on AC calculations rather than DC calculations. Varactor diodes may also be rated according to the amount of AC

input power that they can accept or the amount of AC input power that they can produce when used in specific applications.

Several symbols that are commonly used to represent the varactor diode are shown in Figure 4-7. The symbol on the left actually includes a small symbol of a capacitor thus making the device easy to identify.

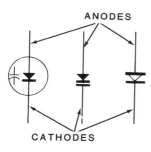

Figure 4-7 Commonly used varactor diode symbols.

Applications

Varactor diodes are used to replace conventional capacitors in many applications. A varactor diode is often used to vary the frequency of a resonant circuit. In this application the diode may represent the only capacitance in the circuit or represent only part of the capacitance of the circuit. The frequency of the resonant circuit is controlled by simply varying the diode's reverse bias voltage. This means that the diode must be connected so that it is properly biased within the circuit. A basic circuit arrangement is shown in Figure 4-8. Capacitor C_1 is quite large and therefore has a very low reactance at the resonant frequency of the parallel resonant circuit which consists of the varactor diode and the inductor L.

Capacitor C_1 is used only to prevent the DC bias from flowing through the inductor. The DC voltage, resistor R, and capacitor C_2 are used to reverse bias the diode as shown. The capacitance of the varactor diode changes as the bias voltage is varied and therefore changes the frequency of the resonant cir-

cuit. When used in this manner the varactor diode becomes a tuning component and is often called a tuning varactor or voltage variable capacitor.

Varactor diodes may also be used in many other applications. For example, they find extensive use in various types of high frequency amplifiers and in devices known as frequency multipliers. They are also used in the automatic frequency control (AFC) circuits found in many FM radios and they are used in a variety of circuits that are found in both AM receivers and transmitters.

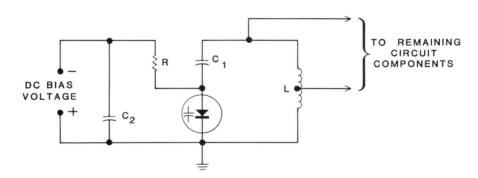

Figure 4-8 Controlling frequency with a varactor diode.

11. The varactor diode is used in applications which utilize its internal _____.

12. A varactor diode is usually operated with a reverse bias voltage that is lower than its _____ rating.

13. When the reverse voltage across a varactor diode increases, the depletion region widens and its internal capacitance _____.

14. When a varactor diode is forward biased with a voltage that exceeds its internal _____, the device cannot be effectively used as a capacitor.

15. The amount of capacitance change that occurs in a varactor diode as a result of changes in temperature is expressed as a _____.

16. The Q of a varactor diode is generally very high.

    A. True
    B. False

17. A varactor diode has no internal resistance as far as AC signals are concerned.

    A. True
    B. False

18. The Q of a varactor will _____ as the reverse bias voltage across the device increases.

19. The frequency required to reduce a varactor diode's Q to a value of 1 is referred to as the _____ frequency.

20. In order to have a high Q, a varactor diode must have a very low _____ current.

21. Most varactor diodes are made from _____.

22. A varactor diode is often used as a _____ component in a resonant circuit.

High Frequency Diodes

There are a variety of special diode types designed for use in high frequency applications. These include both switching (on-off) types and radio frequency signal generating types. All of these diodes operate in the VHF, UHF, and microwave regions. In this section you will examine some of these diodes. The most important types include the PIN, IMPATT, hot carrier, and Gunn effect diodes.

PIN Diodes

All of the diodes previously described utilize a PN junction which is formed by doping a single semiconductor material with different impurities. Each of these diodes contain just two oppositely doped sections which meet to form a single junction. However, another special type of diode is constructed in a slightly different manner. This diode contains an undoped or intrinsic (pure semiconductor) region which is sandwiched between heavily doped N and P sections as shown in Figure 4-9. Since this diode contains a P-type layer, an intrinsic or I layer, and an N-type layer, it is commonly referred to as a PIN diode.

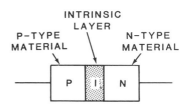

Figure 4-9 PIN diode construction.

PIN diodes are capable of changing from one operating state to another at an extremely fast rate and are therefore used in high-speed switching and pulse generation applications. When used as a switch, the PIN diode is forced to turn on or off by subjecting it to forward or reverse bias voltages.

When a PIN diode is forward-biased, it's forward resistance will change considerably as its forward voltage is varied. Although most other diodes exhibit a variable resistance characteristic, they do not have the wide resistance range of the PIN diode. Furthermore, the internal resistance of the PIN diode changes linearly with forward voltage and the device responds well to low bias voltages and currents.

The resistance of a typical PIN diode may change from 10,000 ohms to 1 ohm when its forward current changes from 0.001 to 100 milliamperes. This characteristic makes the device suitable for use in certain types of limiter circuits or they may simply be used as attenuators (current-controlled resistors). In general, PIN diodes find their greatest applications at the higher (microwave) frequencies. Both low power and high power devices are available. These devices are often packaged like the varactor diodes shown in Figure 4-6.

These diodes are packaged in hermetically sealed metal cases which serve as heat sinks. Each case also provides an electrical connection to one end of the diode within the case (the cathode or anode end) and the protruding metal lead provides the other diode connection.

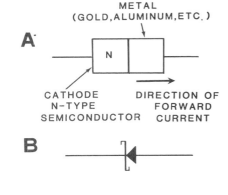

Figure 4-10 Hermetically sealed PIN diode packages.
(Courtesy of Hewlett Packard)

Schottky Diodes

The Schottky diode or Schottky barrier is formed by placing an N-type semiconductor material (usually silicon) in contact with a metal such as gold, silver or aluminum to form a metal-to-semiconductor junction. This diode operates in a manner similar to ordinary PN junction diodes but there are several important differences. The barrier voltage developed within the device is approximately one half as great as the barrier voltage within an ordinary silicon diode. This means that the forward voltage drop across the diode is approximately 0.3 volts instead of 0.6 or 0.7 volts. Also, the Schottky diode operates with majority carriers (electrons); virtually no minority carriers are involved. This means that the reverse or leakage current through the device is extremely small. Figure 4-11 shows the basic Schottky diode construction and the symbol used to represent it.

Figure 4-11 A Schottky diode and its schematic symbol.

Schottky diodes are sometimes called hot carrier diodes (HCD). The name hot carrier diode is used because, in a forward-biased HCD, the electron possesses a high level of kinetic energy as it moves from the N-type material across the junction to the metallic anode.

The Schottky diode is able to change operating states (turn on and off) much faster than ordinary PN junction diodes, and it is used extensively to process high frequency AC signals. This device finds extensive use in microwave electronic mixers (circuits which combine AC signals), detectors (circuits which use rectification as a means of extracting information from AC signals), and high speed digital logic circuits. Figure 4-12 shows some typical diodes for use in microwave circuits.

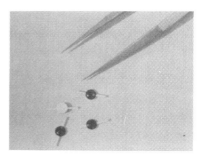

Figure 4-12 Typical Schottky barrier diode
packages.
(Courtesy of Hewlett Packard)

velocity. As the voltage is increased toward a threshold voltage, current and the velocity of the electrons also increase. If the voltage is increased beyond threshold, the device begins to exhibit negative resistance as a result of a decrease in electron velocity and current. The electrons gain so much energy that they actually change to a higher, ordinarily empty, energy band. This phenomenon is the reason that the term **transferred electron effect** is sometimes used to describe the Gunn effect-electrons which have been transferred from the conduction band, to a higher energy band, where their movement is slowed.

Gunn-Effect Devices

Gunn-effect diodes are capable of producing oscillations when they are used in conjunction with a resonant circuit and a DC operating voltage. These diodes are usually made of N-type gallium arsenide (GaAs) semiconductor crystals and do not have a PN junction like an ordinary diode. However, these devices display a negative resistance characteristic within their bulk semiconductor material.

Application of a voltage across a slice of N-type GaAs produces an electric field that causes current to flow toward the positive end. It is a characteristic of N-type GaAs that electrons move through the material at high

When electrons have been transferred to this higher energy band, they enter the cathode more rapidly than they leave, resulting in the gathering of electrons at the cathode end in an accumulation called a **domain**. The domain produces an electric field concentrated at the cathode end and falling off to a level below threshold at the anode end. As shown in Figure 4-13A through F, the domain drifts across the device to the anode and leaves as a pulse, and the cycle repeats (the X axis in Figure 4-13 represents the thickness of the Gunn diode). As a consequence, the output from the Gunn diode is a series of current pulses whose frequency is determined by the width of the drift zone. These pulses can be used to generate microwave signals when they are applied to a resonant cavity (a resonant cavity is essentially a tuned LC circuit that is capable of oscillating at microwave frequencies).

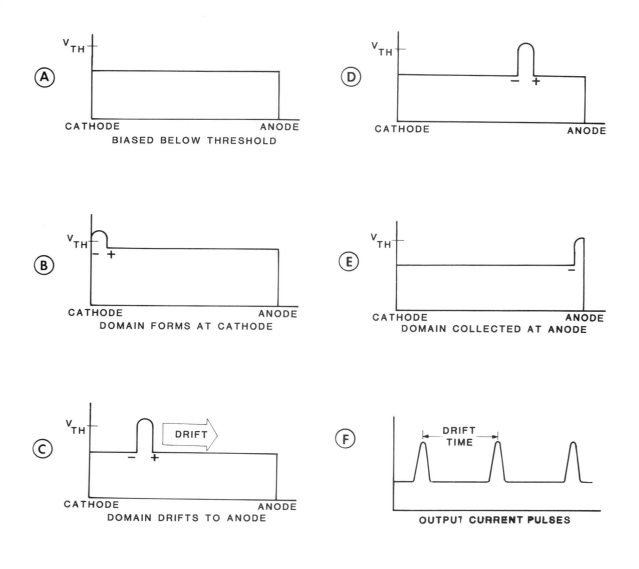

Figure 4-13 Gunn Diode domain drift.

IMPATT Diodes

The construction of an IMPATT (Impact Avalanche Transit Time) diode is similar to that of a zener diode and like a zener diode, an IMPATT diode can withstand avalanche current. Figure 4-14 shows a PN junction that is reverse biased with a voltage sufficient to cause the diode to be in avalanche breakdown. During avalanche breakdown, a small increase in applied voltage causes a large increase in current. If an external resonant circuit applies a high frequency RF voltage to the reverse biased junction, the RF voltage alternately aids and opposes the reverse bias. As a consequence, avalanche current is alternately aided and opposed. However, avalanche current changes are not instantaneous because, as indicated in the Figure 4-14, a large number of electrons move from the avalanche zone into the drift zone. As a result, a phase difference is introduced between the applied voltage changes and avalanche current changes. If the RF current and RF voltage are out of phase by more than 90 degrees, the junction is exhibiting negative resistance to the RF signal. This is because, as the RF voltage increases, the RF current decreases and vice versa.

In a practical application, IMPATT diodes are reverse biased just above the avalanche breakdown point, V_B in Figure 4-15A. This allows the positive transitions of the applied RF voltage to reverse bias the diode into avalanche. As shown in Figure 4-15B, avalanche current then builds up and peaks

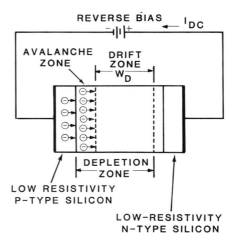

Figure 4-14 IMPATT diode input and output phase relationships.

when the applied voltage returns to zero volts at 180 degrees through its cycle. As the applied voltage goes through its negative half cycle, avalanche decreases, producing a sharply peaked current pulse as shown in Figure 4-15C. This pulse then drifts toward the positively charged side of the junction at an almost constant speed.

The time required for this pulse to pass through the drift zone is determined by the width of the drift zone. The frequency of the output pulse is determined by controlling the width of the drift zone. Therefore, the frequency of the output pulse is determined by controlling the width of the drift zone at the time of manufacture.

As the current pulse drifts through the diode, it induces a positive square wave of current into the external resonant circuit, as indicated in Figure 4-15D. This square wave is present during the negative half cycle of the applied RF signal. Therefore, it is 180 degrees out of phase with the input signal. Thus, the diode generates RF energy as a result of negative resistance.

IMPATT diodes are primarily used to produce microwave radio frequencies above 3000 MHz and are capable of producing higher power levels than Gunn diodes. Silicon, gallium arsenide and, occasionally, germanium are used in the manufacture of IMPATT diodes.

A disadvantage of IMPATT diodes is that they produce higher noise levels than Gunn diodes. However, noise can be minimized by using GaAs IMPATT diodes and proper circuit design. Another disadvantage of IMPATT diodes is that they require relatively high operating voltage.

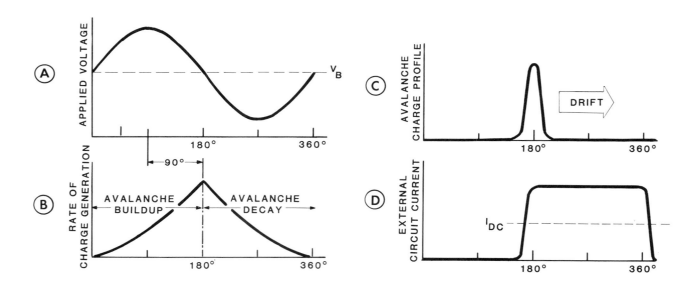

Figure 4-15 IMPATT diode waveforms.

23. A diode which contains a P-type material, an intrinsic layer, and an N-type material is referred to as a _____ diode.

24. The diode packages shown in Figure 4-6 are typical of those devices that are designed to operate at _____ frequencies.

25. IMPATT diodes provide a means of generating _____ _____ _____.

26. The Schottky diode utilizes a metal-to-semiconductor junction.

   A. True
   B. False

27. The barrier voltage within a Schottky HCD is approximately equal to _____ volts.

28. Gunn-effect diodes and IMPATT diodes are capable of generating RF power without the use of external components or voltages.

   A. True
   B. False

29. A special type of PIN diode which exhibits an abrupt change in internal resistance is called a _____ _____ diode.

30. IMPATT diodes operate in the reverse biased condition.

   A. True
   B. False

31. To forward bias a Schottky diode, the N-type semiconductor element is made _____ with respect to the metallic element.

Modification of the construction and/or doping of the basic PN junction makes it possible to produce semiconductor diodes with characteristics that make them useful for special applications.

The PN junction of a tunnel diode is heavily doped. This results in a high internal barrier voltage and a very narrow depletion region, as well as a low reverse breakdown voltage. One of the most significant characteristics of tunnel diodes is the negative resistance region of their operating curves. This negative resistance characteristic makes tunnel diodes useful as a part of the microwave oscillator circuit.

A certain amount of capacitance is present in all PN junction diodes when they are reverse biased. This capacitance results from the depletion region acting as an insulator between the two conductive regions of the diode. Varactor diodes are specially constructed to make use of this capacitance. Decreasing the reverse bias voltage on a varactor diode causes the capacitance of the diode to increase. Consequently, varactor diodes are also called voltage variable capacitors. Varactor diodes are frequently used to control the frequency of resonant circuits.

In a PIN diode, the N-type region and the P-type region are separated by a layer of intrinsic (I) material. A reverse-biased PIN diode does not conduct and acts like a capacitor. When forward bias is applied, the diode acts as a low series resistance. One of the most frequent applications of PIN diodes is as an electronic switch.

Schottky diodes consist of an N-type material in contact with a metal. This construction results in an internal barrier voltage that is approximately one half as great as the barrier voltage in an ordinary PN junction. These devices conduct primarily by means of electrons, with practically no hole current. As a consequence, leakage current through a reverse biased Schottky diode is extremely small. Schottky diodes are used in microwave mixer and detector circuits and in high speed logic circuits. Schottky diodes are also known as Schottky-barrier diodes and Hot Carrier Diodes (HCD).

Gunn diodes have a negative resistance characteristic similar to that of a tunnel diode even though both devices operate on a different principle. The operation of a Gunn diode does not depend on a PN junction or contact characteristics. N-type gallium arsenide is the most commonly used material in the fabrication of Gunn diodes. However, several other materials that display the Gunn effect are also used. Accumulations of electrons called domains are produced by an electric field within the device. As a result, Gunn diodes produce a pulsed current output. The primary application of Gunn diodes is in microwave oscillator circuits.

IMPATT diodes, like Gunn diodes, produce a pulsed current output. For this reason, they are also used as part of microwave oscillator circuits. However, IMPATT diodes are capable of producing higher output power than Gunn Diodes. IMPATT diodes make use of a PN junction that is reverse biased to cause the diode to operate as an avalanche breakdown device. This enables the negative resistance of the device to introduce a 180 degree phase shift between the input signal and the output current pulses. IMPATT diodes are usually used as part of a microwave oscillator circuit. They are capable of producing higher RF output power than other diodes. However, they also produce more noise. IMPATT diodes can be made of gallium arsenide, silicon, and less frequently, germanium.

Unit 5
Bipolar Transistor Operation

Contents

Introduction

This unit on bipolar transistor operation, is designed to familiarize you with the bipolar transistor and its basic principle of operation. You will examine some of the basic construction techniques used to form these important components and you will learn how transistors are used to amplify electronic signals. You will also learn a simple method of testing transistors to determine if they are defective.

A transistor is a three element electronic solid state component used to control electron flow. The amount of current flowing through a transistor can be controlled by varying the proper polarity voltages to the three elements. The 3 elements or leads of a transistor are the base, emitter, and collector. By controlling the current, useful applications such as amplification, oscillation, and switching can be achieved.

Examine your unit objectives closely and relate them to the text material as you read the unit. When you can satisfy all of the unit objectives, you will have successfully completed this unit.

When you have completed this unit on bipolar transistors you should be able to:

1. Describe the basic physical construction of the two basic types of bipolar transistors.

2. Understand the basic principle behind bipolar transistor operation.

3. Show how a bipolar transistor should be biased for normal operation.

4. Explain the relationship between emitter current, base current, and collector current in a bipolar transistor.

5. Identify the three basic transistor amplifier circuit arrangements or configurations.

6. Test bipolar transistors with an ohmmeter to determine if they are shorted or open.

7. Determine if a transistor is a PNP or NPN type using only an ohmmeter.

8. Determine which lead is the base lead using only an ohmmeter.

9. From schematic diagrams identify:

   NPN transistors
   PNP transistors
   Base, emitter, and collector leads
   The 3 circuit configurations

PNP And NPN Configurations

You are now ready to examine a solid-state component that is related to the PN junction diode. This device is called a bipolar transistor although it is sometimes referred to as a junction transistor, BJT, or simply a transistor. Throughout this unit the words transistor and bipolar transistor have the same meaning and are used interchangeably. Later in this course you will examine another basic type of transistor known as a unipolar (UJT) transistor. The terms bipolar and unipolar relate to the manner in which transistors operate and transistors in general are classified either as bipolar or unipolar devices.

A bipolar transistor is constructed from germanium and silicon semiconductor materials just like a PN junction diode, however, this device utilizes three alternately doped semiconductor regions instead of two as in the case of the diode. These three semiconductor regions may be arranged in two different ways. One arrangement is shown in Figure 5-1A. Notice that an N-type semiconductor layer is sandwiched between two P-type semiconductor layers to form what is commonly referred to as a PNP type transistor. The middle or N region is called the base of the transistor and the outer or P regions are called the emitter and collector of the transistor. Normally, the base region is much thinner than the emitter and collector regions and it is also lightly doped in relation to the other two. The appropriate leads must also be attached as shown to provide electrical connections to the base, emitter, and collector leads.

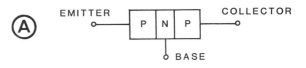

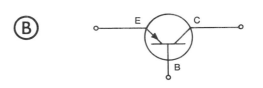

Figure 5-1 A basic PNP transistor and its symbol.

The PNP transistor is usually represented in circuit diagrams or schematics by the symbol shown in Figure 5-1B. The emitter, base, and collector of the transistor are identified by the letters E, B and C respectively; however, these letters may not always appear with the symbol. It is also important to remember that the arrow is always located in the emitter lead and it always points to the N-type material.

The second method of arranging the semiconductor layers in a bipolar transistor is shown in Figure 5-2A. Notice that this time a layer of P-type material is sandwiched between two layers of N-type material to form what is commonly known as an NPN transistor. Like the PNP transistor this device utilizes a narrow middle region which is referred to as the base. The outer layers are again referred to as the emitter and collector regions.

The NPN transistor is usually represented by the symbol shown in Figure 5-2B. The respective emitter, base, and collector leads may or may not be identified by the additional letters (E, B and C) as shown. Notice that the only difference between the PNP and NPN transistor symbols is the direction of the arrow which represents the emitter of the device. The arrow points toward the base in the PNP symbol but away from the base in the NPN symbol.

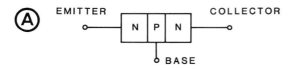

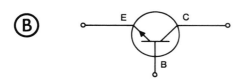

Figure 5-2 A basic NPN transistor and its symbol.

Because PNP and NPN transistors have three alternately doped semiconductor regions, these devices have two junctions of P and N type materials. In other words each transistor has a PN junction between its emitter and base regions and a PN junction between its base and collector regions. Therefore, bipolar transistors have three layers and two junctions as opposed to PN junction diodes which have two layers and only one junction.

Transistor Construction

Figures 5-1A and 5-2A show the basic PNP and NPN structures, but still do not show exactly how these devices are constructed. The three semiconductor layers are not just simply pressed together as suggested by these simple illustrations. Instead, the three layers are often formed by using construction techniques that are similar to those used to construct ordinary PN junction diodes. A complete analysis of all of the various construction techniques would be impossible at this time since many different processes have been devised and these processes are constantly being improved. Therefore, we will only briefly review a few basic techniques from which many of the newer techniques have been developed.

Construction Techniques

In general, bipolar transistors are constructed by using the alloyed method or the diffusion method or by using variations of these basic processes. Figure 5-3A shows a transistor that has been formed by the alloyed method. The three regions are formed by placing indium pellets on each side of a thin N-type semiconductor crystal. The entire unit is then heated until the pellets melt and diffuse into the N-type material. After the heat is removed, the indium (a trivalent material) and the N-type material recrystallize to form a PNP transistor as shown.

Figure 5-3B shows a transistor that has been formed by the diffusion process. This transistor is formed by subjecting an N-type semiconductor crystal to a trivalent impurity at an elevated temperature. The trivalent impurity is allowed to penetrate or diffuse into the crystal to form a P-type region. Then the process is repeated with a pentavalent impurity to form a small N-type region within the P-type region previously formed. This results in an NPN transistor as shown; however, PNP transistors can be formed using the same basic process.

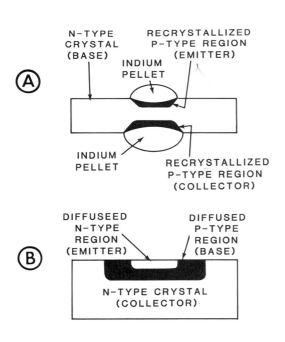

Figure 5-3 Cross-sections of typical bipolar transistors.

Both of the construction techniques just described are similar to the techniques used to construct PN junction diodes. Each technique offers certain advantages and disadvantages and both techniques may be combined in various ways to produce transistors with certain desirable characteristics. Also the physical size and shape of the layers is important in determining how the transistor will operate. In general, the base is usually smaller than the collector region. The reasons for this relationship are explained later in this unit.

In addition to the two basic construction methods just described, more refined methods are often used to construct transistors. For example, some transistors are formed through an epitaxial growth process. This type of transistor is formed by growing a crystalline film on top of a basic semiconductor crystal. This film has the same crystalline structure as the original material and can contain either type of impurity. This process makes it possible to grow additional P-type or N-type layers to obtain the necessary PNP or NPN arrangement.

Transistors may also be constructed by using the epitaxial growth process along with the diffusion process previously mentioned. The general shape of a transistor is also an important factor. The transistor shown in Figure 5-3B is formed by selectively diffusing the base and emitter regions into the collector region so that the device has a flat top or in other words all of the regions lie in a single plane. When this type of construction is used, the device is referred to as a planar transistor. However, in many cases a trench or moat is etched around the base and emitter regions of the transistor which leaves the base and emitter standing on a sort of plateau or mesa. This technique is used to expose the base or emitter so that electrical connections can be easily made to these regions. This type of construction is also used to define the areas where the regions meet. Figure 5-4 shows a transistor that utilizes this mesa type of construction as well as epitaxially grown and diffused regions. The construction starts by epitaxially growing a lightly doped P-type layer (the base) on top of an N-type semiconductor crystal (the collector). Then the emitter is formed by diffusing an N-type region into the epitaxial layer. Finally, the area around the emitter is etched away to form a mesa structure.

It is important to realize that a large number of transistors are simultaneously formed with any given process. Generally, the transistors are formed on a very thin semiconductor wafer or disc which may be as large

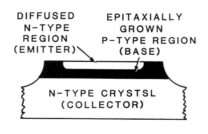

DIFFUSED N-TYPE REGION (EMITTER)

EPITAXIALLY GROWN P-TYPE REGION (BASE)

N-TYPE CRYSTSL (COLLECTOR)

Figure 5-4 Cross-section of an Epitaxial-Base transistor which utilizes a mesa structure.

as 1 1/2 to 2 inches in diameter. After all of the regions are simultaneously and selectively formed, the wafer is then scribed and broken into many hundreds or thousands of transistors.

Packaging

Once a transistor is constructed, it must be installed in a suitable container or package which will protect the device and provide a means of making electrical connections to its emitter, base, and collector regions. The package may also serve as a heat sink which will drain heat away from the transistor and prevent it from being damaged. Many different package or case styles are available to cover a wide range of possible applications. Several commonly used transistor packages are shown in Figure 5-5. Packages A through D are designed for low power applications where either small electrical signals or currents are to be amplified or controlled.

The transistors are actually located inside of these packages and their emitter, base and collector regions are electrically connected to the leads which protrude from the bottom of each package. Packages A through C are made of metal and the transistors are hermetically sealed within these units to protect them from humidity, dust, and other materials which might contaminate them. The transistor's collector region is often connected directly to the metal case; therefore, one of the leads may serve as the collector lead as well as a case connection.

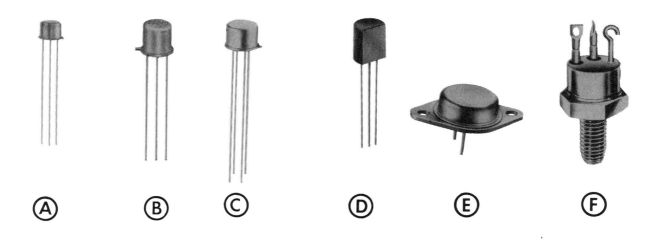

Figure 5-5 Typical low power and high power transistor packages.
(*Courtesy of United Corporation*)

When it is necessary to have the transistor electrically insulated from the metal case, but still have an electrical connection to the case, package C may be used. With this arrangement, a fourth lead is provided which serves as a case connection.

Package D is made of plastic but similar epoxy types are also available. The transistor is encapsulated within the plastic housing and the three protruding leads connect to the transistor's emitter, base, and collector regions. In general, plastic or epoxy transistors are not as expensive as the metal types, however the plastic units are usually not as rugged as comparable metal devices. All of the low power units just discussed have parallel leads which are closely spaced. These components are designed so

that their leads can be inserted through the eyelets in printed circuit boards and then soldered. However, the leads can also be shortened so that these devices can be plugged into matching sockets or when necessary the leads can simply be spread apart and soldered to adjacent components.

Packages E and F are designed for high power applications. These packages are made of metal and are larger than the low power units. The transistors are usually connected within these packages so that their collectors are securely attached to the metal cases. The heavy cases are then able to extract or drain away the heat that is generated within the transistor. These packages are also designed so that they can be securely fastened to a metal chassis or frame which can in turn extract the heat which accumulates within them. The external chassis or frame serves as a heat sink and effectively increases the power handling ability of the device. The two pins protruding from the bottom of package E serve as the emitter and base connections while the package case serves as the collector connection. The case of package F serves as the collector connection, however an additional collector lead (the hooked one) is also provided at the top of the package along with the emitter and base leads.

Transistor packages are usually referred to by a reference designation that indicates its size and configuration. The letters TO (transistor outline) followed by a number is the commonly used package identifier. The common packages are TO-5 (Figure 5-5A), TO-18 (Figure 5-5B), TO-39 (Figure 5-5C), TO-92 (Figure 5-5D), TO-3 or TO-66 (Figure 5-5E) and TO-59 or TO-63 (Figure 5-5F).

Although all of the packages shown in Figure 5-5 are used to support, protect, and improve the operation of transistors, these same packages may also be used with other types of solid-state components. In fact many of the solid-state components that you will examine in the following units are mounted in packages which are either identical or similar to the ones shown here. In addition to these packages, there are many other types that are commonly used.

Due to the tremendous assortment of packages available, it is difficult to form absolute rules for identifying the various leads or terminals on each device. Each type of package has its particular lead arrangement and it is usually best to refer to the manufacturer's applicable specifications to identify the leads or terminals on each device.

1. A bipolar transistor which utilizes an N-type region that is sandwiched between two P-type regions is referred to as a ___P N P___ transistor.

2. A bipolar transistor which utilizes a P-type region which is sandwiched between two N-type regions is referred to as a/an ___N P N___ transistor.

3. Bipolar transistors have only one PN junction.

   A. True
   B. False

4. Bipolar transistors are formed by simply pressing together three alternately doped layers of germanium or silicon.

   A. True
   B. False

5. Because its diffused regions lie in a single plane, the transistor shown in Figure 5-3B is referred to as a _____ . _ transistor.

6. Since the emitter and base regions of the transistor shown in Figure 5-4 rest on a plateau, the device utilizes a _____ type of construction.

7. Packages A,B,C, and D in Figure 5-5 are designed for low power applications.

   A. True
   B. False

8. The collectors of transistors mounted in packages E and F of Figure 5-5 are connected to the package cases so that the _____ generated within the device may be quickly removed.

9. The arrow in the schematic symbol is always located in the _____ lead and always points to the _____ type material.

10. Bipolar transistors can only be manufactured by using the diffusion process.

    A. True
    B. False

Although bipolar transistors may be used in various ways, their basic and most important function is to provide amplification. In other words bipolar transistors are used primarily to boost the strength or amplitude of electronic signals. The signals applied to a transistor and the amplified signals obtained from the transistor can be expressed in terms of voltage, current, or power. However, you will soon discover that the bipolar transistor is actually controlled by the current flowing through its terminals and this current is, in turn, controlled by varying the input voltage. Furthermore, the transistor must be properly biased by external voltages so that its emitter, base, and collector regions interact in the desired manner.

Biasing NPN Transistors

Like the PN junction diode that you examined in a previous unit, the bipolar transistor must be properly biased in order to perform a useful function. However, the bipolar transistor has two PN junctions and both of these junctions must be properly biased if the device is to function properly. For example, consider the NPN transistor shown in Figure 5-6. Notice that an external voltage has been applied to the base and collector regions. The junction formed between these two regions is commonly referred to as the collector-base junction or simply the collector junction. The collector and base regions effectively form a PN junction diode even though the base is very thin and lightly

doped. The external voltage is being used to reverse bias the collector junction as shown. The reverse-biased collector junction will function in a manner similar to a reverse-biased PN junction diode and conduct only a very small leakage current. This extremely small current is supported by minority carriers in the N-type collector and P-type base regions. As explained in a previous unit, the minority carriers are the holes in the N-type material and the electrons in the P-type material. In this case the number of minority carriers in both regions (the base in particular) is very small and the resulting leakage current is therefore quite small. Under normal operating conditions, the collector junction of an NPN transistor is reverse biased by connecting a positive supply voltage to its collector lead.

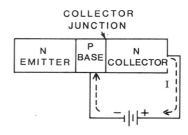

Figure 5-6 Reverse biased collector junction.

Now we will remove the external voltage shown in Figure 5-6 and apply it to the emitter and base regions of the NPN transistor as shown in Figure 5-7. Although the base region is extremely thin and lightly doped with respect to the emitter, these two regions still function exactly like a PN junction diode. The junction between these two regions is often referred to as the emitter-base junction or simply the emitter junction. The external voltage is being used to forward bias the emitter junction because the N-type emitter and P-type base regions are respectively connected to the negative and positive terminals of the voltage source. The majority carriers within the two regions are now forced to combine at the emitter junction; however, the majority carriers (electrons) within the emitter greatly exceed the majority carriers (holes) in the base. This means that the base cannot supply enough holes to combine with all of the electrons that cross the junction. However, a certain amount of forward current will still flow through the emitter junction and the external voltage source. Under normal operating conditions the emitter junction must be forward-biased in this manner.

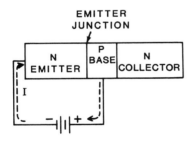

Figure 5-7 Forward biased emitter junction.

We have now seen the action which takes place in an NPN transistor when the collector junction and the emitter junction of the

device are independently subjected to the proper bias voltages. However, our analysis of transistor action is not yet complete. This is because both junctions of the transistor must be simultaneously biased if the device is to operate properly. In other words, the emitter junction must be forward-biased and the collector junction must be reverse-biased as shown in Figure 5-8. When both junctions are biased in this manner, the action that takes place in the transistor is quite different.

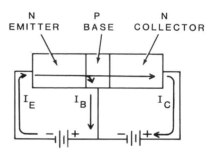

Figure 5-8 A properly biased NPN transistor showing internal and external current flow.

We will now consider the action that takes place in the properly biased NPN transistor shown in Figure 5-8. Since the emitter junction is forward biased, electrons in the N-type emitter and the holes in the P-type base (the majority carriers) are forced to move toward the emitter junction and combine. However this portion of the transistor can no longer function like an independent diode and conduct a forward current which flows only through the emitter junction and its external bias voltage source. In fact, most of the current that flows into the emitter of the transistor will now flow completely through the transistor's base and collector regions. To understand how this action occurs we must consider both the physical and electrical characteristics of the transistor's emitter, base, and collector regions.

The transistor has a very thin base region which is lightly doped with respect to the emitter. This means that the majority carriers (electrons in an NPN transistor) in the emitter will greatly exceed the majority carriers (holes) in the base. Therefore, most of the electrons that cross the emitter junction do not combine with holes and tend to accumulate in the base region. However, the electrons that are inserted or injected into the narrow base region by the emitter are now influenced by the positive potential applied to the collector region. In fact, most of these electrons are swept across the collector junction and into the collector region and from there into the positive side of the external voltage source which is used to reverse bias the collector junction. This action is shown in Figure 5-9. Typically, 95 to 99 percent of the electrons supplied by the emitter flow through the collector region and into the external voltage source. This current is referred to as collector current and is usually designated as I_C as shown in Figure 5-8. The remaining emitter injected electrons (1 to 5 percent) combine with holes in the base region and therefore support a small current which flows out of the base region. This relatively small current is generally referred to as base current and is

designated as I_B. The current that flows into the emitter region is generally referred to as emitter current and is designated as I_E. The relationship between I_E, I_B, and I_C can be stated mathematically with the following equation:

$$I_E = I_B + I_C$$

The equation simply states that the emitter current (I_E) is equal to the sum of the base current (I_B) and the collector current (I_C). Another way of looking at this is that the collector current is equal to the emitter current less the current lost to the base.

$$I_C = I_E - I_B$$

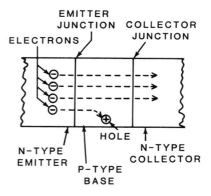

Figure 5-9 NPN transistor action.

If the base region of the transistor was not extremely thin, the action just described could not occur. The thin base region makes it possible for the emitter injected electrons to move quickly into the collector region. A large base region would minimize the interaction between the emitter and collector regions and the transistor would act more like two separately biased diodes as shown in Figures 5-6 and 5-7. Furthermore, the emitter is doped much more heavily than the base region so that a large number of electrons will be supplied to the base and be subsequently pulled into the collector region. The collector region is lightly doped although not as lightly doped as the base region and the collector is also considerably larger than either the base or emitter regions. This relationship allows the collector to produce a sufficient number of minority carriers while the number of majority carriers are reduced to a low level. A large number of majority carriers could actually interfere with collector operation because they tend to inhibit the production of minority carriers.

In our analysis of transistor action we must also consider two additional points which are very important. First, it is important to realize that the emitter-base junction of the transistor has characteristics similar to those of a PN junction diode. A barrier voltage is therefore produced across the emitter junction of the device and this voltage must be exceeded before forward current can flow through the junction. Throughout our previous discussion we have assumed that the forward bias voltage exceeds this internal barrier voltage and we have also assumed that the emitter current is held to a safe value. The internal barrier voltage within a particular transistor is determined by the type of semiconductor material used to construct the device. The internal barrier voltage is typically 0.3 volts for germanium transistors and 0.7 volts for silicon transistors. This means that transistors, like PN junction diodes, will exhibit a relatively low voltage drop (0.3 or 0.7 volts) across their emitter junctions under normal operating conditions. A second important point to consider is that the collector-base junction of the transistor must be subjected to a positive potential that is high enough to attract most of the electrons supplied by the emitter. Therefore, the reverse bias voltage applied to the collector-base junction is usually much higher than the forward bias voltage across the emitter-base junction.

Biasing PNP Transistors

Up to this point we have considered only the action which takes place in a properly biased NPN transistor. The only difference between the NPN and PNP transistors is the type of charge carriers involved and the polarity of the external bias voltages. The NPN transistor utilizes electrons as charge carriers but the PNP transistor utilizes holes. In other words, NPN transistors depend on the movement of electrons between their emitter, base, and collector regions while the operation of PNP devices is explained by the movement of holes. Also, since the structures of PNP and NPN transistors are opposite, these devices require external bias voltages which are also opposite. This means that the external bias voltages shown in Figure 5-8 must be reversed so that the PNP transistor is biased as shown in Figure 5-10. If you examine this figure closely you will see that the emitter junction is forward-biased and the collector junction is reverse-biased. Figure 5-10 also shows that the emitter, base, and collector currents flowing through the PNP device are opposite to those flowing through the NPN device.

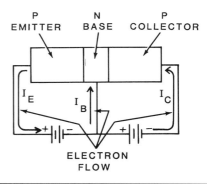

Figure 5-10 A properly biased PNP transistor showing external current flow.

11. A bipolar transistor has _____ PN junctions.

12. The junction formed between the emitter and base regions is called the _____ junction.

13. The _____ junction in a bipolar transistor is usually reverse-biased.

14. The _____ junction in a bipolar transistor is usually forward-biased.

15. The collector current flowing through the NPN transistor shown in Figure 5-8 is exactly equal to the emitter current.

    A. True
    B. False

16. The base current in a bipolar transistor will usually be equal to only 1 to 5 percent of the emitter current.

    A. True
    B. False

17. All of the current flowing through a transistor flows through the _____ lead.

Transistor Amplification

Now that we have seen the basic action that takes place in a properly biased transistor we will go one step further and see how a transistor can be used to amplify electronic signals. We will continue our analysis of the NPN transistor shown in Figure 5-8; however, we will replace the transistor with its electronic symbol and therefore, represent the circuit as shown in Figure 5-11. Notice that the external voltage sources that are used to forward bias the emitter junction and reverse bias the collector junction have also been labeled V_{EE} and V_{CC} respectively. These two symbols are widely accepted as standard designations for the operating voltages used in this type of circuit. Furthermore, we have made V_{EE} variable so that we may vary the forward bias voltage applied to the emitter junction. The transistor has also been designated as Q_1 since the Q designation is widely used to represent transistors in electronic circuits.

In the previous discussion on transistor action we saw that the emitter current (I_E), base current (I_B), and collector current (I_C) could be related mathematically by the expression $I_E = I_B + I_C$. We assumed that all three currents were constant. However, all three of these currents vary in a proportional manner as indicated by the mathematical expression previously stated. For example, if I_E doubled in value, then I_B and I_C would likewise double in value. The reason for this is quite simple and can be easily explained. The number of electrons that the emitter supplies to the base region is determined by the amount of forward bias voltage across the emitter junction. When the forward bias voltage increases (V_{EE} in-creases) more electrons are injected into the base region (I_E increases), therefore more electrons are swept into the collector region to become collector current (I_C increases). The additional electrons that do not move into the collector region combine with holes in the base region and become base current (I_B increases). When the forward bias voltage decreases, the opposite effect is produced and all three currents decrease. The three currents are directly proportional. In other words, they all vary by the same percentage.

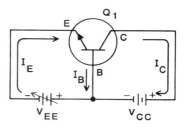

Figure 5-11 Schematic representation of a properly biased NPN transistor.

In order for the transistor shown in Figure 5-11 to provide amplification, the device must be capable of accepting an input signal (current or voltage) and provide an output signal that is greater in strength or amplitude. The transistor cannot perform this function if it remains connected as shown in Figure 5-11. Instead we must make several changes to the circuit so that it appears as shown in Figure 5-12. Notice that we have added a resistor between the collector of transistor Q_1 and the positive terminal of

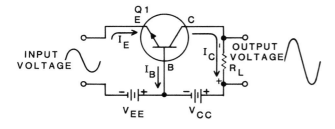

Figure 5-12 A basic NPN transistor amplifier circuit.

V_{CC}. This component is commonly referred to as a load resistor and is labeled R_L. This resistor is used to develop an output voltage with the polarity indicated. In other words the collector current must flow through this resistor to produce a specific voltage drop. We have also separated the connection between the emitter of transistor Q_1 and the negative terminal of V_{EE}. These open connections serve as the input of the amplifier circuit and allow an input signal to be inserted between the emitter of Q_1 and V_{EE}. Therefore the input is not an open circuit, but is completed when an external voltage source is connected between the emitter and V_{EE}. Furthermore, the values of V_{EE} and V_{CC} have been selected to bias transistor Q_1 so that the values of I_E, I_B and I_C are high enough to permit proper circuit operation.

The emitter junction of transistor Q_1 is forward-biased and like a PN junction diode, has a relatively low resistance. However, the collector junction is reverse-biased and therefore has a relatively high resistance. In spite of the tremendous difference in emitter junction resistance and collector junction resistance, the emitter current

(I_E) is almost equal to the collector current (I_C). The value of I_C is only slightly less than I_E because a small portion of I_E flows out of the base region to become base current (I_B). The load resistor can therefore have a high value without greatly restricting the value of I_C.

If a small voltage is applied to the input terminals which aids the forward bias voltage (V_{EE}), the value of V_{EE} is effectively increased. This will cause I_E to increase and I_C will also increase. This will increase the voltage drop across the load resistor (R_L). However, when the input voltage is reversed so that it opposes V_{EE}, the value of V_{EE} is effectively reduced and both I_E and I_C will decrease. This will cause the voltage drop across R_L to decrease. A change in input voltage can therefore produce a corresponding change in output voltage. However, the output voltage change will be much greater than the input change. This is because the output voltage is developed across a high output load resistance (R_L) and the input voltage is applied to the low resistance offered by the emitter junction. Thus, a very low input voltage can control the value of I_E which, in turn, determines the value of I_C. Even though I_C remains slightly less than I_E, it is forced to flow through a higher resistance and produces a higher output voltage.

The action just described will occur if the input signal is either a DC or an AC voltage. The important point to remember is that any change in input voltage is greatly amplified by the circuit so that a much larger but proportional change in output voltage is obtained. It is also important to realize that the input signal is not simply increased in size, but it is used to control the conduction of transistor Q_1 and this transistor in turn controls the current through the load resistor (R_L). The transistor is made to take energy from the external power source V_{CC} and apply it to the load resistor in the form of an output voltage whose value is controlled by the small input voltage.

The transistor shown in Figure 5-12 is being used to convert a relatively low voltage to a much higher voltage. Any circuit which performs this basic function is commonly referred to as a voltage amplifier. Such circuits are widely used in electronic equipment since it is often necessary to raise both DC and AC voltages to higher levels so that they may be effectively used. Later in this unit you will also discover that the transistor can be connected in a different circuit configuration which will allow it to amplify currents as well as voltages. Any circuit which is used for the sole purpose of converting a low current to a higher current is commonly referred to as a current amplifier. In some textbooks the terms current and power amplifiers are interchangeable.

Although an NPN transistor was used in the amplifier circuit shown in Figure 5-12, a PNP transistor will perform the same basic function. However; in the PNP circuit the bias voltages (V_{EE} and V_{CC}) must be reversed and the respective currents (I_E, I_B, and I_C) will flow in the opposite directions.

18. The emitter, base, and collector currents in a bipolar transistor tend to vary by the same percentage.

    A. True
    B. False

19. PNP transistor bias polarities are the same as the NPN transistor.

    A. True
    B. False

20. The transistor shown in Figure 5-12 is being used as a _____ amplifier.

21. If an input voltage is applied to the circuit shown in Figure 5-12 that aids the forward bias voltage (V_{EE}), the voltage across R_L will _____.

22. Only NPN transistors may be used in the amplifier circuit shown in Figure 5-12, regardless of the polarities of V_{EE} and V_{CC}.

    A. True
    B. False

23. For proper operation a transistor's emitter-base junction must be _____ biased and its base-collector junction must be _____ biased.

Transistor Circuit Arrangements

As explained previously, the bipolar transistor is used primarily as an amplifying device. However, there is more than one way of using a transistor to provide amplification and each method offers certain advantages or benefits but also has certain disadvantages or limitations. Basically, the bipolar transistor may be utilized in three different circuit arrangements (configurations) to perform the amplifying function.

In each arrangement one of the transistor's three leads is used as a common reference point and the other two leads serve as input and output connections. The three circuit arrangements are referred to as the common-base circuit, the common-emitter circuit, and the common-collector circuit, and each arrangement may be constructed by using either NPN or PNP transistors.

In many cases the common leads are connected to circuit or chassis ground and the circuit arrangements are then referred to as grounded-base, grounded-emitter, and grounded-collector circuits. Furthermore, in each circuit arrangement the transistor's emitter junction is always forward-biased while the collector junction is reverse-biased. We will now briefly examine each of these basic circuit arrangements to demonstrate the flexibility of the bipolar transistor. In the next unit of instruction these circuits will again be described but in much greater detail.

Common Base Circuits

In the common-base circuit, the transistor's

base region is used as a common reference point and the emitter and collector regions serve as the input and output connections. This basic circuit arrangement is shown in Figure 5-13. Figure 5-13A shows how an NPN transistor is connected in the common-base configuration while Figure 5-13B shows how the same arrangement is formed with a PNP transistor. Notice that both circuits are arranged in the same basic manner, but the polarity of the bias voltages (V_{EE} and V_{CC}) are opposite so that the NPN and PNP transistors are properly biased. This common-base arrangement was also described in the previous sections of this unit although the previous circuits were not identified as common-base circuits. This arrangement was used in the previous discussions because it serves as an excellent model for explaining basic transistor operation.

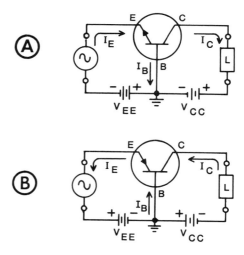

Figure 5-13 Common-base circuits.

In the common-base circuit the input signal is applied between the transistor's emitter and base and the output signal appears between the transistor's collector and base. The base is therefore common to both the input and output. The input signal applied to the circuit is represented by the AC generator symbol. The output voltage is usually developed across a resistive component, but in some cases, components which have inductive, as well as resistive properties (coils, relays, or motors), may be used. In general, the component that is connected to the output of the circuit is referred to as a **load**. In Figure 5-13 the load is identified by the box which contains the letter "L".

The common-base circuit is useful because (as explained earlier) it provides voltage amplification. However, this circuit has other characteristics which should be considered. Since the emitter junction of the transistor is forward-biased the input signal sees a very low emitter-to-base resistance or in other words the transistor has a low input resistance. However, the collector junction of the transistor is reverse-biased and therefore offers a relatively high resistance. The resistance across the output terminals (collector-to-base) of the transistor is therefore much higher than the input resistance. The input resistance of a typical low power transistor that is connected in the common-base configuration could be as low as 30 or 40 ohms; however, the output resistance of the transistor could be as high as 1 megohm.

The common-base circuit provides slightly less output current than input current because I_C is always slightly lower than I_E. In other words there is a slight current loss between the input and output terminals of the circuit. However, the circuit is still useful as a voltage amplifier as explained earlier. Since the input and output currents are almost the same and because the output voltage developed across the load can be much higher than the input signal voltage, the output power produced by the circuit is much higher than the input power applied to the circuit. Remember that power is equal to the current times the voltage or $P = I \times E$. The common-base circuit therefore provides power amplification as well as voltage amplification.

Common Emitter Circuits

Bipolar transistors (NPN and PNP) may also be connected as shown in Figure 5-14. Notice that the input signal is applied to the base and referenced to the emitter and the output signal is developed between the collector and emitter. Since the emitter of each transistor is common to the input and output of each circuit, the arrangements are referred to as common-emitter circuits. Notice that the emitter junction of each transistor is still forward-biased while the collector junction of each device is reverse-biased.

Also, the NPN circuit shown in Figure 5-14A and the PNP circuit shown in Figure 5-14B require bias voltages that are opposite in polarity. Notice that the forward bias voltage (designated as V_{BB}) is now applied to the base of the transistor instead of the emitter. In other words the forward bias voltage (V_{BB}) is applied so that it controls the base current (I_B) instead of the emitter current (I_E). Only the base current flows through bias voltage V_{BB} and the AC generator (signal source). The reverse bias voltage (V_{CC}) is applied through the output load to the collector and emitter regions of the transistor and only the collector current (I_C) flows through the load and V_{CC}.

Although it may not be apparent, V_{CC} does actually reverse bias the collector-base junction of the transistor. Figure 5-15 shows how this reverse bias is obtained in an NPN common-emitter circuit. If V_{CC} is connected across the emitter and collector regions of the transistor as shown, the emitter-base junction becomes forward-biased and therefore has a low resistance. However, V_{CC} causes the N-type collector to be positive with respect to the P-type base and the collector-base junction is therefore reverse-biased. The forward-biased emitter-base junction effectively allows the negative terminal of V_{CC} to be applied to the P-type base region. V_{CC} is effectively applied to the collector and base regions as shown. If V_{CC} was reversed, the collector junction would become forward-biased and the emitter junction would become reverse-biased; however, this condition would be undesirable. The same basic action also occurs in a PNP transistor even though the PNP device requires bias voltage polarities which are exactly opposite to those used with an NPN device.

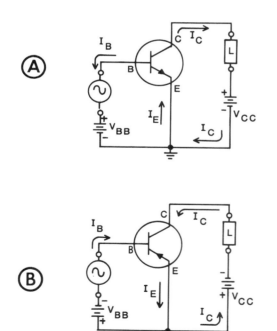

Figure 5-14 Common-emitter circuits.

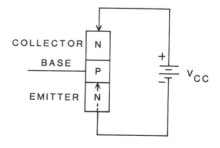

Figure 5-15 The reverse-biased collector junction in an NPN common-emitter circuit.

The three currents (I_E, I_B, and I_C) in the common-emitter circuit still have the same relationship as before and still vary in a proportional manner. In other words I_E is still equal to $I_B + I_C$ even though I_B has now become the input current and I_C has become the output current. Any change in the input current (I_B) will result in a proportional change in I_E and I_C. The difference between this circuit and the common-base circuit previously described is simply one of reference. We have now connected the transistor so that its very low base current is being used to control its much larger collector current.

When an input signal voltage is applied to the common-emitter circuit, it either aids or opposes V_{BB} and causes I_B to either increase or decrease. The much higher emitter and collector currents (which are almost equal) are forced to increase and decrease by the same percentage. Since a very low I_B is used to control a relatively high I_C, the transistor is now being used to provide current amplification as well as voltage amplification. This of course means that the power supplied to the load is much greater than the power applied to the input of the circuit. In fact, the common-emitter arrangement, when compared to either the common-base or common-collector circuit arrangements, provides a higher power output for a given input power. The common-emitter configuration is the only circuit configuration that provides voltage, current, and power amplification. The common-emitter arrangement is also the only circuit configuration that is capable of 180 degree phase shift between the input and output signals.

A transistor, that is used in a common-emitter arrangement will also have a low input resistance but this resistance will not be as low as the input resistance of a common-base arrangement. This is because the input base current for the common-emitter circuit is much lower than the input emitter current of the common-base circuit. Also, the output resistance of a transistor connected in a common-emitter arrangement will be somewhat lower than when it is connected in a common-base circuit. A typical low power transistor might have an input resistance of 1000 and 2000 ohms and an output resistance of 50,000 or 60,000 ohms.

Common Collector Circuits

The NPN and PNP transistors shown in Figure 5-16 are connected so that the collector of each transistor is used as a common reference point. The base and the emitter regions serve as the input and output connections. These circuits are known as common-collector circuits.

In the common-collector circuit the input signal voltage is applied between the base and collector regions of the transistor. The input either aids or opposes the transistor's forward bias and causes I_B to vary accordingly. This in turn causes I_E and I_C to vary by the same percentage. An output voltage is developed across the load which is connected between the emitter and collector regions of the transistor. The emitter-current (I_E) flowing through the load is much greater than the base current (I_B). Thus the circuit provides an increase in current between its input and output terminals. However, the voltage developed across the load will always be slightly lower than the voltage applied to the circuit. The slightly lower voltage will appear at the emitter of the transistor because the device tends to maintain a relatively constant voltage drop across its emitter-base junction. This forward voltage drop may be equal to approximately 0.3 volts if the transistor is made of germanium or 0.7 volts if the transistor is made of silicon.

The output voltage appearing at the emitter of the transistor tends to track or follow the input voltage applied to the transistor's base. For this reason the common-collector circuit is often called an emitter-follower.

The common-collector circuit functions as a current amplifier but does not produce an increase in voltage. However, the increase in output current (even though the output voltage is slightly less than the input voltage) results in a moderate increase in power. Also, the input resistance of any transistor connected in a common-collector arrangement is extremely high. This is because the input resistance is that resistance which appears across the reverse-biased collector-base junction. This input resistance can be as high as several hundred thousand ohms in a typical low power transistor. However, the output resistance appearing between the transistor's emitter and collector regions will be much lower (often as low as several hundred ohms) because of the relatively high emitter current (I_E) that flows through the output lead.

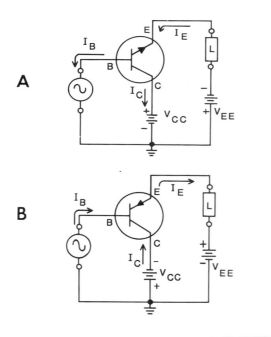

Figure 5-16 Common-collector circuits.

The common-collector circuit is not useful as a voltage amplifying device. Instead, this circuit is widely used in applications where its high input resistance and low output resistance can perform a useful function. The circuit is often used to couple high impedance sources to low impedance loads and therefore perform the same basic function as an impedance matching transformer. This is sometimes referred to as an isolation circuit or buffer circuit. However the high current gain does provide a substantial amount of power gain.

The bias voltages (V_{BB}, V_{EE}, and V_{CC}) used in the three circuit arrangements just described may appear to restrict or interfere with the flow of signal current through these circuits; however, this does not occur. The bias voltages are effectively shorted as far as the input and output signals are concerned, although in some applications large capacitors are placed across the voltage sources to insure that they offer minimum impedance to signal currents. For example, a large electrolytic capacitor could be placed across V_{CC} in Figure 5-16 to ensure that the collector of the transistor is at ground potential as far as AC signals are concerned.

24. A transistor that is connected in a common-base arrangement provides a substantial current gain.

    A. True
    B. False

25. Regardless of the type of circuit arrangement used, the transistor's emitter junction must always be _____- biased.

26. The component connected to the output of a common-base, common-emitter, or common-collector circuit is commonly referred to as a _____.

27. The common-base circuit provides both voltage and power amplification.

    A. True
    B. False

28. The transistor used in a common-base circuit will have a low input _____.

29. The common-emitter circuit provides an increase in current, voltage, and power.

    A. True
    B. False

30. In a common-emitter circuit I_B serves as the _____ current.

31. The common-_____ circuit arrangement provides the highest power amplification.

32. In the common-collector circuit, the input signal is applied between the _____ and _____ regions of the transistor.

33. The common-collector circuit does not provide voltage amplification.

    A. True
    B. False

34. The V_{BE} drop across a silicon transistor used in the common-emitter configuration will be _____ V.

35. Another name for the common-collector circuit arrangement is the emitter-_____.

36. The circuit arrangement preferred for impedance matching a high impedance source to a low impedance load is the common-_____ arrangement.

Testing Bipolar Transistors

Although bipolar transistors are solid-state devices which are capable of operating for extremely long periods of time without failure, these devices do occasionally become defective. The failure of a particular transistor can be caused by excessively high temperatures, currents or voltages, or by subjecting the component to extreme mechanical stress. As a result of these electrical or mechanical abuses, the transistor may short or open internally or in some cases only the characteristics of the device may be altered.

Test Procedures

Test equipment is readily available for testing transistors either in or out of an electronic circuit. These instruments will indicate when a transistor is open or shorted, how well it amplifies, and whether or not the device is passing an excessive amount of undesirable leakage current. A simple transistor tester is shown in Figure 5-17. However, the most common troubles (opens and shorts) can be easily detected with an ohmmeter.

The ohmmeter may be used to determine if a short or open exists between the transistor's emitter and base, base and collector, or emitter and collector. When checking a transistor with an ohmmeter, you are simply looking for unusually low (near zero) or unusually high (nearly infinite) resistances. Since a bipolar transistor has two PN junctions, it may be compared with two diodes that are connected back-to-back. Each diode (PN junction) will exhibit a low resistance

Figure 5-17 A simple bipolar transistor and junction diode tester that checks for shorts, opens, gain and leakage.

when forward-biased and a high resistance when reverse-biased. The ohmmeter battery is the source of the forward and reverse bias voltage. In order to test a transistor we must therefore check the forward and reverse resistance of each junction.

Testing NPN Transistors

To check the forward resistance of each junction in an NPN transistor, the ohmmeter leads must be connected as shown in Figure 5-18A. The positive (normally red) lead is connected to the base while the negative (usually black) lead is connected first to the transistor's emitter and then to its collector. The ohmmeter will send a forward current first through the emitter base junction, then through the base-collector junction and should indicate that both junctions have a relatively low resistance (typically several hundred ohms or less). Next, the ohmmeter leads are reversed and both junctions are again checked as shown in Figure 5-18B. This time both junctions are reverse-biased and the ohmmeter should indicate that both junctions have a relatively high resistance (several hundred thousand ohms or greater).

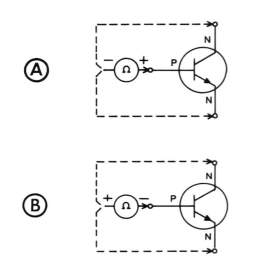

Figure 5-18 Testing an NPN transistor.

Testing PNP Transistors

When checking a PNP transistor the same basic procedure is followed; however, the ohmmeter connections are exactly opposite to those shown in Figure 5-18. The proper ohmmeter connections for checking the forward resistance of the two junctions in a PNP device are shown in Figure 5-19A and the proper connections for checking reverse resistance is shown in Figure 5-19B.

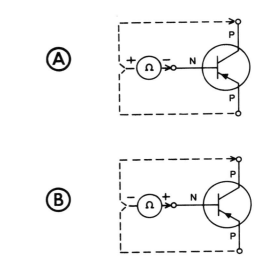

Figure 5-19 Testing a PNP transistor.

Although the two test procedures just described are quite simple, several general precautions should be observed when the tests are performed. When checking the forward resistance of the junctions, you should use a low resistance range that will allow the meter to present a mid-range indication or one that is easy to read. When checking the reverse resistance, a higher range must be used to obtain a convenient reading. Furthermore, the specific resistance readings that you obtain have no real meaning. You are interested only in determining that each PN junction has a high ratio or forward and reverse resistance. However, through experience you will find that these forward and reverse resistances generally fall within certain ranges. Several hundred ohms of forward resistance and several hundred thousand ohms to several megohms of reverse resistance is typical for both low and medium power transistors, but high power transistors generally have forward and reverse resistances that are somewhat lower. Also, transistors that are made from silicon usually have higher forward and reverse resistances than transistors that are made from germanium. A good rule of thumb for both transistors and signal diodes is at least a 1 to 100 ratio front to back.

When checking low power transistors you should avoid using the lowest or highest resistance ranges on your ohmmeter if possible, because your ohmmeter will supply its maximum current or voltage to the transistor when these ranges are used. In most cases this will not damage a transistor but it is still a good practice to use.

When a junction has very low forward and reverse resistances (particularly when both resistances are equal) the junction is effectively shorted and the transistor is defective. When a junction has an extremely high resistance in both the forward and reverse directions, it is effectively open and the transistor is again considered to be defective. Both the forward and reverse junction resistances must be measured before the true condition of the junction is revealed.

CAUTION NOTE

All meters can NOT be used to statically check a PN junction. Some DMM (digital multi meters) don't supply enough power to cause the junction to allow current to flow.

When you check transistors with an ohmmeter, you may notice that the forward and reverse resistances change when different resistance ranges are used. This occurs because the ohmmeter applies a different voltage and current to the transistor on each resistance range. In addition, resistance readings often vary considerably from one ohmmeter to another. One way of increasing the reliability of PN junction resistance measurements is to compare the resistance readings of the transistor under test with resistance readings from a known good transistor of the same type and using the same test equipment.

Identifying A Transistor

Sometimes you may find it necessary to determine if a given transistor is an NPN or PNP device. This identification can also be performed with an ohmmeter. As in testing a transistor with an ohmmeter, you must know the polarity of the voltage at the ohmmeter leads. Normally, the internal ohmmeter battery will be connected so that a positive potential will appear at the red or plus lead of the ohmmeter and a negative potential appears at the black or minus lead.

In some ohmmeters the color of the leads will not be red and black in which case you should check your unit by referring to the ohmmeter's circuit diagram or schematic.

In order to forward bias a PN junction, you must apply a bias voltage to it so that the cathode (N) is negative and the anode (P) is positive. You must reverse the polarity to reverse bias the junction. With this information and the knowledge of your ohmmeter operation, you can identify a transistor as being PNP or NPN.

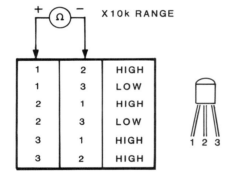

Figure 5-20 Identifing the unknown transistor.

An ohmmeter set to midrange was used to develop the information in Figure 5-20.

The transistor used was a 417-201 (X29A829) and without consideration for which lead was the base, emitter, or collector the readings were plotted in Figure 5-20.

Note that when lead 1 is (+) and lead 3 is (−) the meter reading is low. Also when lead 2 is (+) and lead 3 is (−) the reading is low. The low reading indicates that the transistor was forward biased or conducting. Since lead 3 is in both low readings, lead 3 is the base lead and since forward bias was achieved when a negative was applied to the base, the base must be N-type material. Therefore, the 417-201 transistor must be a properly operating PNP transistor.

Desk-Top Experiment 3
Identifying a Transistor

Introduction

The purpose of this exercise is to use the information that you have gained to analyze Figure 5-21 and determine as much information as possible about the transistor in question.

Objectives

To determine which leads are the base, collector, and emitter.

To determine the type of transistor and if it is operational or defective.

Procedure

The transistor in Figure 5-21 has a V_{BE} of 0.3 volt. With that piece of information and the data contained in the figure determine:

1. Determine the type of transistor.

2. Label the base, collector, and emitter leads.

   Base _____ .
   Collector _____ .
   Emitter _____ .

3. Explain why you decided which lead was which.

   Base

   Collector

   Emitter

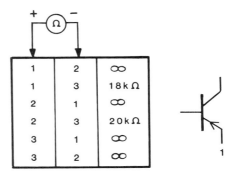

Figure 5-21 Data for Desk-Top Experiment 3.

4. Using the transistor in Figure 5-21 draw a properly biased amplifier arrangement and label as many points as you can.

5. On the circuit you just drew indicate the direction of current flow.

6. Explain current flow as it relates to your circuit.

   Current flow is

Discussion

From the schematic symbol you should have identified the transistor as a bipolar PNP type, and that lead 1 is the emitter and lead 3 is the base.

Because of the value of V_{BE} you should have realized that the transistor was a germanium transistor and that the majority carriers are holes.

Using the information in the chart you should have decided that lead 3 is the base and that proper bias for the transistor is positive to the collector and negative to the base. You should also have decided that the transistor was operational.

Your amplifier arrangement could have been any of the arrangements discussed up to this point in the text. On your schematic drawing you should have indicated that current flow was from the emitter to the collector.

Summary

The first thing to catch your eye should have been the V_{BE} value of 0.3 volts. This is the voltage drop across a germanium barrier junction. Thus, the transistor was manufactured from the solid state semiconductor material germanium (Ge).

Next, in Figure 5-21 the arrow is in lead number 1. The arrow is always in the emitter lead and it always points to the N-type material. Therefore, the transistor must be a PNP type and pin 3 must be the base lead. If you missed the arrow in the schematic symbol the base could have been identified from the information contained in the chart. Notice in the chart, that lead 3 is in both low readings and that it is negative in both cases. If you remembered that the emitter-base junction must be forward biased, then, you would have realized that if a negative voltage forward biased the base, then the base must be N-type material.

Since a transistor is a 3 element device, once the emitter and base are identified as leads 1 and 3 respectfully, then, by the process of elimination lead 2 must be the collector.

The transistor is assumed to be operational since it does not measure either shorted or open.

The majority current carriers in a bipolar PNP transistor are holes. Holes have a positive charge and move from the emitter to the reverse biased collector. Since the collector is P-type material its supply voltage must be negative.

37. Bipolar transistors may be tested for opens or shorts by using an _____.

38. In testing a bipolar transistor you may observe that the reverse resistance of the emitter-base junction is low. Therefore, the junction is:

    A. open
    B. shorted
    C. good
    D. cannot be determined

39. In testing a bipolar transistor you observe that the reverse resistance of the base-collector junction is high. Therefore, the junction is:

    A. open
    B. shorted
    C. good
    D. cannot be determined

40. From the information in Figure 5-22, determine if the transistor under test is:

    good/defective
    NPN/PNP
    base lead is pin_____.

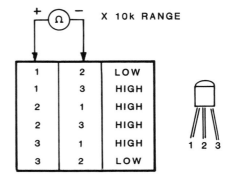

Figure 5-22 Data for question 40.

Unit Summary

Bipolar transistors are two junction devices that consist of three alternately doped semiconductor regions. A transistor that has a P-type region between two N-type regions is called an NPN transistor. Similarly, a transistor having an N-type region between two P-type regions is a PNP transistor.

The center region of a transistor is called the base and the two outer regions are called the emitter and the collector. The base region is usually much thinner than the other two regions and is lightly doped. The base is P-type material in an NPN transistor and N-type material in a PNP transistor.

Several methods are used in the construction of transistors. The three most commonly used methods are the alloyed method, the diffusion method, and epitaxial growth. Many transistors are constructed simultaneously on a single semiconducor wafer. The wafer is then broken up into individual components. Each component is enclosed in a package to protect the device from mechanical damage and the environment, as well as making it possible to connect the transistor to other circuit components. Transistors are enclosed in many different types of packages, depending on their intended application.

The junction formed between the transistor's emitter and base is often called the emitter-base or emitter junction. Similarly, the junction between the collector and base is called the collector-base or collector junction. Under normal operating conditions, the emitter junction is forward biased and the collector junction is reverse biased. This means that, in an NPN transistor, the N-type emitter must be negative with respect to the P-type base, while the base is negative with respect to the N-type collector.

In a properly biased NPN transistor, electrons flow from the transistor emitter to the base. A few of these electrons leave the base as base current. However, most of the electrons pass through the base to the positive collector as collector current. Therefore, the emitter current is equal to the sum of the base and collector currents. However, when transistor circuits are being analyzed, emitter current and collector current can be considered equal because the base current is so low. Remember that in N-type material electrons are the majority carriers.

Operation of a PNP transistor is similar to that of an NPN transistor. Since holes are the majority carriers in P-type material the operation of a PNP transistor depends on hole movement instead of electrons. Holes are confined to the internal structure of the transistor. Emitter, base, and collector currents are still produced by electrons flowing into and out of the device. A PNP transistor is properly biased when its emitter is positive with respect to the base and the collector is negative with respect to the base.

Practically all transistor applications are based on a transistor's ability to amplify weak electronic signals. Amplification is the ability to increase the amplitude of an electronic signal. The amplitude of the input signal can be expressed in terms of voltage or current. However, before the transistor can amplify, it must be connected to an external circuit in a manner that allows the input signal to control the transistor's conduction. The transistor then controls the current through an external load, such as a resistor. A transistor amplifier circuit may be any one of three basic circuit configurations, common emitter, common base, or common collector. In a given configuration, one of the transistor leads, emitter, base, or collector, is the circuit reference point while the other two leads serve as input and output connections.

In a common emitter circuit, the emitter of the transistor is the common reference point while the input is applied to the base and the output is taken from the collector. The common emitter circuit amplifies voltage, current, and power. This circuit configuration has low input resistance and high output resistance and is the only configuration capable of shifting the input signal by 180 degrees. The common emitter configuration is the most popular of the three configurations.

When a transistor is connected in the common base configuration, the base of the transistor serves as the common reference point. The input voltage is applied to the emitter and the output appears across the collector load resistance. A common base circuit amplifies voltage and power. However, output current is slightly less than input current because collector current is always less than emitter current. The common base configuration is the least used of the three configurations.

The collector of the transistor serves as the reference point in a common collector circuit. The input is applied to the base and the output is taken from a load resistance connected to the emitter. Common collector circuits amplify current and power. However, the output voltage is always less than the input voltage. This is because part of the input is dropped across the emitter-base junction of the transistor. Common collector circuits cannot be used in applications where voltage amplification is required. However, the high input resistance and low output resistance of this configuration makes the common collector circuit useful in applications where impedance matching is required. The common collector circuit is frequently called an emitter follower because the emitter output voltage is approximately the same as the input voltage, or it follows the input voltage.

Transistors can be easily tested for opens or shorts with most ohmmeters. The ohmmeter should indicate that the forward resistance of each transistor junction is low and that the reverse resistance is high. If a given junction has a high resistance in both directions, the junction is probably open. Conversely, if the ohmmeter indicates that the junction resistance is low in both directions, the junction is probably shorted. The ohmmeter can also be used to determine if an unknown transistor is an NPN or PNP type. Remember that the emitter-base junction must be forward biased and the base-collector junction must be reverse biased for the transistor to operate properly.

Unit 6
Bipolar Transistor Characteristics

Contents

Introduction

In the previous unit you learned how the bipolar transistor operates and how the device can be used to provide amplification of electronic signals. In this unit you will examine in greater detail some of the bipolar transistor's important electrical characteristics when it is used in each of the three basic amplifier configurations. As explained previously these three arrangements are referred to as the common-base circuit, the common-emitter circuit, and the common-collector circuit. Examine your unit objectives closely to determine exactly what you are to learn during your study of this unit.

When you have completed this unit on bipolar transistor characteristics you should be able to:

1. Determine the approximate current gain (alpha) of a common-base transistor by using the transistor's collector characteristic curves.

2. Explain the meaning of the term "alpha cutoff frequency."

3. Explain what I_{CBO} is and why it is important.

4. Determine the current gain (beta) of a common-emitter transistor by using the transistor's collector characteristic curves.

5. Explain the meaning of the term "beta cutoff frequency."

6. Explain what I_{CEO} is and why it is important.

7. Determine alpha when beta is known and vice-versa.

8. Determine the approximate input and output resistance of a common-collector transistor circuit.

9. Explain why a common emitter arrangement inverts.

10. Explain the term buffer amplifier.

11. List the circuit arrangement/configuration preferred for the following applications, and explain why it is the preferred choice.

    power amplification
    voltage amplification
    current amplification
    polarity inversion
    impedance matching
    isolation
    high frequency operation

Characteristics Of Common-Base Circuits

As explained in the previous unit, a bipolar transistor may be connected in the common-base configuration to provide voltage and power amplification but not current amplification. When a transistor is used in a common-base circuit, an input signal voltage is inserted between the transistor's emitter and base. The input voltage effectively varies the transistor's emitter current which in turn causes the transistor's base and collector currents to vary in a proportional manner. The transistor's emitter and collector currents serve as input and output currents respectively and since the collector current is slightly lower than the emitter current, the common-base circuit cannot provide current amplification. However, a load resistor can be connected between the transistor's collector and base so that the transistor's collector current will flow through the load and develop an output voltage. The voltage developed across this load resistor will be much higher than the input signal voltage thus allowing the circuit to provide voltage amplification. Due to the higher voltage developed across the load resistance, the power supplied to the load will greatly exceed the input power required to operate the circuit, thus, allowing the circuit to provide power amplification.

The common-base circuit represents just one of three basic circuit arrangements in which the bipolar transistor can be used to provide amplification. The common-base arrangement is not widely used in electronic equipment but it is necessary to examine certain common-base characteristics to obtain an overall understanding of

transistor operation. We will now examine some of these important characteristics.

Collector Characteristic Curves

When a bipolar transistor is connected in the common-base configuration, the transistor's collector current may be controlled by varying the emitter current flowing through the device. However, the transistor's collector current may also be controlled (to a lesser degree) by varying the reverse bias voltage applied to the collector junction. Furthermore, the transistor's collector current tends to vary in direct proportion to changes in emitter current, but the collector current does not always vary proportionally with changes in reverse bias voltage. Therefore, in order to effectively show the current-voltage relationships that exist within the transistor, it is necessary to plot the related current and voltage values on a graph.

When plotted, the curve formed shows how the transistor's collector current, emitter current, and reverse voltage are related. This curve shows the transistor's important electrical characteristics when it is connected in the common-base configuration and is referred to as a characteristic curve.

To adequately show the transistor's electrical characteristics it is necessary to plot a number of characteristic curves. Each curve is plotted for a specific value of emitter current starting at zero and working up to a certain maximum value. In each case the emitter current is held constant while the reverse bias voltage across the collector junction is varied and the change in collector current is observed. The corresponding values are then plotted on a graph and a group of curves are formed.

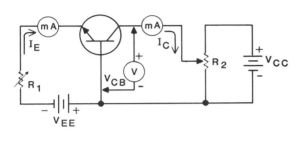

Figure 6-1 Circuits used to determine characteristic curves for a common-base configuration.

The characteristic curves for a common-base configuration may be obtained by using a circuit like the one shown in Figure 6-1. The transistor used in this circuit is an NPN device; however, the same circuit could be used to plot the curves for a PNP transistor if the polarities of the bias voltages were reversed. Notice that a variable resistor (R_1) is in series with the transistor's emitter and the external voltage source (V_{EE}) that is used to forward bias the transistor's emitter-base junction. The resistor is used to control the transistor's emitter current (I_E). The transistor's collector-base junction is reverse-biased by external voltage source V_{CC} but this voltage is made variable by potentiometer R_2. This potentiometer may be adjusted

to control the reverse voltage applied to the transistor. Two milliammeters are also used in the circuit to measure the emitter current (I_E) and collector current (I_E). Also, a voltmeter is used to measure the reverse voltage across the transistor's collector-to-base junction. This reverse voltage is commonly designated as V_{CB} as shown.

A typical set of characteristic curves are shown in Figure 6-2. To obtain these curves R_1 is adjusted to various values of emitter current (I_E). For each value of emitter current, the transistor's reverse voltage (V_{CB}) is varied over a wide range of values and the transistor's corresponding collector current (I_C) is observed at each value of V_{CB}. When the corresponding V_{CB} and I_C values are graphically plotted for each I_E value, a set of characteristic curves (often called a family of characteristic curves) are obtained. A typical set of these curves are shown in Figure 6-2.

Notice that the characteristic curves in Figure 6-2 are plotted for I_E values which range from 0 to 7 milliamperes. The V_C values are plotted horizontally on the graph and the I_C values are plotted vertically. The curve for $I_E = 0$ is produced when no emitter current flows through the transistor. At this time only a small leakage current can flow through the collector-to-base junction. Notice that this current increases only a small amount as V_{CB} is varied from 0 to more than 50 volts and for all practical purposes is nearly equal to I_E (which is equal to zero). Also notice that the collector-to-base junction must be slightly forward

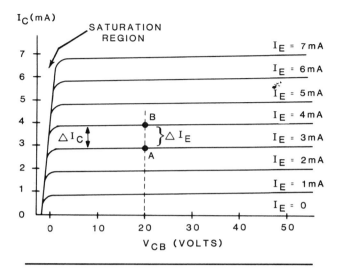

$I_C(mA)$

SATURATION REGION

$I_E = 7\,mA$
$I_E = 6\,mA$
$I_E = 5\,mA$
$I_E = 4\,mA$
$I_E = 3\,mA$
$I_E = 2\,mA$
$I_E = 1\,mA$
$I_E = 0$

ΔI_C
ΔI_E
B
A

V_{CB} (VOLTS)

Figure 6-2 Typical collector characteristic curves for a common-base configuration.

biased (as indicated by V_{CB} values to the left of 0 on the horizontal axis of the graph) to completely reduce I_C to zero. This is a peculiar feature of a common-base transistor circuit, and it occurs because of an inherent collector-to-base junction potential which is produced within the transistor.

Notice that the curve for I_E in mA rises rapidly and then quickly levels off as V_{CB} changes from a very low forward bias voltage to a reverse bias voltage of more than 50 volts. This indicates that I_C rises to a maximum value and then remains almost constant even though V_{CB} varies over a wide range. Notice that I_C never quite reaches a value of 1 milliampere. This is because I_C can never be exactly equal to I_E in a common-base circuit since a portion of I_E flows

out of the transistor's base and becomes base current. The I_E mA curve in Figure 6-2 shows that when I_E is equal to 1 milliampere, I_C will also be approximately 1 milliampere over a wide range of V_{CB} values. Only when V_{CB} is reversed so that it slightly forward biases the transistor's collector-to-base junction, can I_C be reduced to zero. The remaining curves in Figure 6-2 follow the same general pattern as the I_E mA curve, but at higher values of current. Notice that all of the curves initially overlap but then each curve reaches a maximum value and then remains essentially constant. In each case the maximum current (maximum I_C value) is always slightly less than the I_E value. The portion of each curve where collector current rises rapidly with a small change in V_{CB} (before the knee of the curve) is referred to as the "saturation region" as indicated in Figure 6-2.

Operation of the transistor within the saturation region is generally avoided. In most circuits, the transistor is biased to operate in the region where I_C is relatively constant while V_{CB} varies (to the right of the knee of each curve). In this region I_C is controlled primarily by I_E (not by V_{CB}).

The curves in Figure 6-2 may be used to determine the value of I_C for a wide range of I_E and V_{CB} values. For example, if V_{CB} equals 20 volts and I_E equals 3 milliamperes, then I_C will be approximately equal to 2.8 milliamperes as shown. Under these conditions the transistor is operating at point A on the characteristic curves as indicated. However, the transistor's operating point can be moved to point B by simply increasing I_E to 4 milliamperes while V_{CB} remains at 20 volts. At this time the value of I_C will be approximately 3.8 milliamperes. From these examples you can see how useful the curves are for predicting how the transistor will operate under various conditions.

Manufacturers of transistors will often provide characteristic curves like those in Figure 6-2 to show the transistor's common-base characteristics. However, manufacturers usually test a number of transistors (of a given type) and produce a set of average or typical curves. When the transistor's collector current and voltages are plotted as shown in Figure 6-2, the resulting curves are often referred to as static collector characteristic curves or simply collector characteristic curves.

The collector characteristic curves shown in Figure 6-2 are for an NPN transistor, but the same type of curves may also be plotted for a PNP transistor. In the case of a PNP transistor, the related values (I_B, I_C, and V_{CB}) are generally considered to be negative instead of positive as shown in Figure 6-2.

Current Gain

The collector characteristic curves just described effectively show the amount of current amplification that a common-base transistor can provide. In other words the collector curves show how much control the transistor's input (emitter) current has over its output (collector) current.

In the previous unit you learned that a common-base circuit cannot actually provide current amplification because its collector current (I_C) is always slightly less than its emitter current (I_E). However, it is still common practice to describe the common-base transistor in terms of its ability to amplify or provide an increase or gain in current. In other words we may consider the transistor to have a specific current gain but this gain must always be slightly less than 1 or unity. This is equivalent to saying that the transistor provides no current gain at all or that it produces a loss in current gain.

The current gain of a common-base transistor is determined by varying the transistor's emitter current and observing the corresponding change in collector current. This is done while the transistor's reverse bias voltage (V_{CB}) is held constant. The current gain of a common-base transistor is often identified by the Greek letter alpha (α) and is expressed mathematically as:

$$\text{current gain} (\alpha) = \frac{\Delta I_c}{\Delta I_E}$$

This equation simply states that the current gain (alpha) is equal to a small change in collector current divided by a small change in emitter current. It is assumed that the reverse-bias voltage across the collector-to-base junction (V_{CB}) remains constant when these measurements are made. Furthermore, the current gain is expressed simply as a ratio of change in the two currents and is not assigned a specific unit of measurement.

The common-base transistor's collector characteristic curve may be used to graphically determine the transistor's current gain. For example, between points A and B on the curves in Figure 6-2, the emitter current (I_E) varies from 3 milliamperes to 4 milliamperes (with V_{CB} constant at 20 volts). The collector current (I_C) therefore changes from approximately 2.8 milliamperes to approximately 3.8 milliamperes. This means that a change in I_E from 4 to 3 or 1 milliampere will produce a change in I_C of 3.8 − 2.8 or 1 milliampere. The current gain of the transistor may therefore be expressed mathematically as:

$$\text{current gain } (\alpha) = \frac{1\text{ milliampere}}{1\text{ milliampere}} = 1$$

In other words, the change in I_E is equal to the change in I_C, and this results in a gain of 1.

The actual gain of the common-base transistor is not quite equal to 1 as indicated by the previous calculations. Our graphical analysis resulted in a gain of 1 because we could not determine the precise values of I_C from the curves in Figure 6-2. Actually, the change in I_C is slightly less than the change in I_E and the gain of the transistor is just slightly less than 1. Typical values of alpha for most bipolar transistors will range from 0.95 to 0.995.

Manufacturers of transistors usually state the alpha value for each type of transistor that they produce. The alpha value given is usually a minimum expected value which is derived under a specific set of operating conditions (often V_{CB} is 5 volts and I_C is 1 milliampere). Under different operating conditions the alpha value could be slightly different since the transistor's I_E, I_C, and V_{CB} values do not maintain an exact relationship over the transistor's useful operating range.

The transistor's alpha is also commonly referred to as the forward current transfer ratio of the device and is represented by the symbol h_{fb}. In fact this designation is becoming increasingly popular while the alpha designation is finding less use.

The alpha or forward current transfer ratio (h_{fb}) just described is an AC measurement which is derived by observing the corresponding changes in I_E and I_C with V_{CB} constant. However, a similar measurement can be made by using fixed values of I_E and I_C. In other words the corresponding values of I_E and I_C (DC or static values) at a specific operating point on the collector curves may be used to calculate the DC alpha value of the transistor. The DC alpha (forward current transfer ratio) of a device is generally represented by the symbol H_{FB}.

Alpha Cutoff Frequency

When a transistor is subjected to a wide range of AC input signals its current gain (alpha) will not remain constant. In general the current gain of the transistor will decrease when the device is subjected to sufficiently high frequencies. When the time of one cycle (period) of the input AC signal approaches the transit time of a charge carrier (the time required for a charge carrier to pass through the device) the gain of the transistor drops rapidly. The frequency at which the gain (alpha) of the transistor drops to 70.7 percent of its low frequency value is called the alpha cutoff frequency and it is often represented by the symbol f_{ab}. This is an important characteristic of common-base transistor circuits and manufacturers of transistors usually specify the f_{ab} for each type of transistor produced. The f_{ab} value of each transistor is generally determined by measuring the current gain of each device at a reference frequency of 1000 hertz and then the frequency is increased until the gain drops to 70.7 percent of its reference value.

Collector-To-Base Leakage Current

The collector characteristic curves in Figure 6-2 show that the value of I_C decreases as I_E decreases. However, when I_E is reduced to zero, I_C will not decrease to an absolute zero value. Instead a small leakage current will still continue to flow through the transistor. In fact an extremely small collector current will flow even when the transistor's collector-to-base junction acts like a reverse-biased diode and allows a small reverse or leakage current to flow through the junction. As in the case of a reverse-biased diode this leakage current is caused by minority carriers in the collector and base regions.

The fact that a small leakage current can flow through a transistor's reverse-biased collector junction, even when the transistor's emitter is open, is an important consideration. This small leakage current is commonly referred to as the transistor's collector-to-base leakage current and is often represented by the symbol I_{CBO}. In a transistor I_{CBO} is important because it combines with and therefore becomes part of the transistor's collector current (I_C). In other words I_{CBO} still flows through the collector junction even when the transistor is properly biased so that its emitter current (I_E) produces a corresponding collector current (I_C). The transistor's collector current is made up of two components. It consists of that portion of I_E which does not flow through the transistor's base to become base current (I_B) and I_{CBO}.

I_{CBO} is affected only slightly by changes in the reverse bias voltage (V_{CB} across the transistor's collector junction). However I_{CBO} is very sensitive to changes in temperature because it is produced by minority carriers. As a general rule I_{CBO} approximately doubles for every 10 degrees centigrade rise in temperature for both silicon and germanium transistors. However, silicon transistors usually exhibit a much lower I_{CBO} value than comparable germanium transistors.

Fortunately I_{CBO} is extremely small in most transistors (often only a few nanoamperes or microamperes). Therefore its effects are minimal in many circuit applications. In certain critical circuits (especially those sensitive to temperature) its effects must be considered. In our previous discussions and in those which follow, the effects of I_{CBO} are considered to be insignificant.

1. The collector characteristic curves for a common-base transistor show the relationship between the transistor's collector current (I_C) and collector-to-base voltage (V_{CB}) for various values of _____ _____.

2. The characteristic curves shown in Figure 6-2 are plotted for an NPN transistor.

    A. True
    B. False

3. The characteristic curves in Figure 6-2 show that when $I_E = 0$, I_C will be almost equal to _____ even though V_{CB} varies over a wide range.

4. The curves in Figure 6-2 show that _____ is controlled primarily by I_E and is affected only slightly by changes in V_{CB}.

5. The portion of each curve where I_C rises rapidly as V_{CB} increases slightly is referred to as the _____ region.

6. Manufacturers usually test only one transistor of a given type before producing a set of collector characteristic curves for that type of transistor.

    A. True
    B. False

7. The current gain of a common-base transistor is identified by the Greek letter _____.

8. A transistor's alpha may be determined graphically by comparing a change in I_E with a corresponding change in I_C while _____ remains constant.

9. A transistor's alpha will always be slightly less than _____.

10. The frequency at which the alpha of a transistor drops to 70.7 percent of its low frequency value is referred to as the transistor's _____ _____.

11. The small leakage current that flows through a transistor's reverse-biased collector junction when the transistor's emitter is open is represented by the symbol _____.

Characteristics Of Common-Emitter Circuits

When a bipolar transistor is connected in a common-emitter configuration, the input signal voltage is inserted between the transistor's base and emitter and effectively varies the transistor's base current which in turn controls the collector current. The base and collector currents serve as input and output currents respectively. Since the collector current is much higher than the base current, the common-emitter circuit provides current amplification. When a load resistor is connected between the transistor's collector and emitter and the collector current is forced to flow through the load, an output voltage is developed across the load which is greater than the input signal voltage. This results in voltage amplification. The circuit also produces a tremendous increase in power because of the increase in current as well as voltage.

The common-emitter circuit is the most widely used transistor circuit configuration in all types of electronic equipment. Since it is an extremely important configuration, a detailed analysis of some of its important electrical characteristics is desirable.

Collector Characteristic Curves

The various current-voltage relationships in a common-emitter circuit can be quickly analyzed by referring to an applicable set of collector characteristic curves. These curves are plotted in much the same way as the collector curves previously described for the common-base arrangements. However, the curves for the common-emitter circuit show the relationship between the transistor's base current, its collector current, and the voltage across its collector and emitter.

The collector characteristic curves for a common-emitter transistor could be determined by using a circuit like the one shown in Figure 6-3. A potentiometer (R_1) is used to adjust the transistor's base current (I_B) to various values and at each value of I_B the second potentiometer (R_3) is adjusted so that the voltage applied to the transistor's collector and emitter (designated as V_{CE}) is

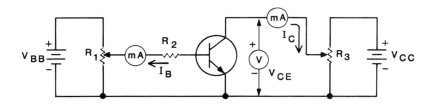

Figure 6-3 Circuit used to determine characteristic curves for a common-emitter transistor.

varied over a wide range. The resulting change in collector current (I_C) is observed in each case. Resistor R_2 is used to limit I_B to a safe value. The related currents and voltages are measured with two milliammeters and one voltmeter as shown.

A typical set of collector characteristic curves for a common-emitter transistor are shown in Figure 6-4. Notice that these curves are plotted for various I_B values which range from 0 to 250 microamperes. The V_{CE} values are plotted horizontally and the I_C values are plotted vertically. The curves are therefore constructed in the same basic manner as the common-base curves previously described. The curve for $I_B = 0$ is produced when no base current flows through the transistor. At this time only a small leakage current flows through the transistor, thus, causing I_C to be almost equal to zero. This leakage current increases only slightly as V_{CE} is increased. The curve for $I_B = 50$ microamperes rises rapidly and quickly levels off as V_{CE} is varied from 0 to more than 50 volts. The collector current (I_C) reaches a maximum value of approximately 0.7 milliamperes as shown because I_C is limited by the amount of emitter current and base current flowing in the transistor. The emitter and base currents in turn depend on the forward bias voltage applied to the transistor's emitter junction.

The remaining curves plotted in Figure 6-4 are for I_B values that range from 75 to 250 microamperes. Each of these curves have the same general shape as the curve for $I_B = 50$ microamperes. However, the remaining curves have a slightly greater slope. The region just to the left of the knee of each curve is referred to as the saturation region. This is the portion of each curve where I_C rises rapidly with an increase in V_{CE}. Normally the transistor is biased so that it will operate beyond the knee of each curve or in other words within the region where I_C changes only slightly with changes in V_{CE}.

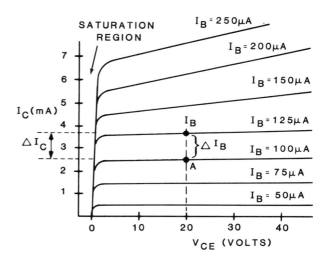

Figure 6-4 Typical collector characteristic curves for a common-emitter transistor.

When used in conjunction with an oscilloscope, this semiconductor curve tracer will automatically plot transistor characteristic curves.

Figure 6-6 shows a set of collector characteristic curves as plotted on an oscilloscope by a semiconductor curve tracer. Here the collector current is plotted as a function of base current and collector-emitter voltage. The sharp upward transition of the collector current shown on the far right represents breakdown in the base-collector junction.

Figure 6-5 Curve tracer.

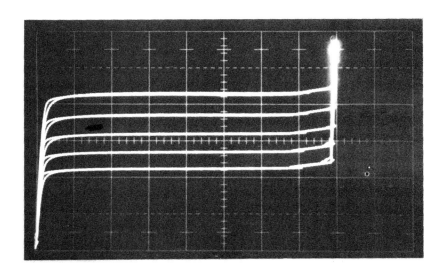

Figure 6-6 Oscilloscope presentation of a set of characteristic curves.

The curves in Figure 6-4 may be used to determine the corresponding values of I_B, I_C, and V_{CE} at any specific operating point. For example, when V_{CE} is equal to 20 volts and I_B is equal to 100 microamperes, I_C will be equal to approximately 2.5 milliamperes as shown. At this time the transistor is operating at point A as shown. The I_C value of 2.5 milliamperes is considerably higher than the I_B value of 100 microamperes. This shows that the common-emitter transistor is capable of providing a substantial current gain.

Current Gain

The amount of current amplification that a common-emitter transistor provides is easily determined from the transistor's collector characteristic curves. The transistor's ability to provide an increase or gain in current is determined by varying the transistor's base current and observing the corresponding change in collector current. However, this is done while the transistor's collector-to-emitter voltage (V_{CE}) is held constant. The current gain of a common-emitter transistor is sometimes identified by the Greek letter beta (β) and is expressed mathematically as:

$$\text{Current gain } (\beta) = \frac{\Delta I_C}{\Delta I_B}$$

This equation states that the current gain is equal to the small change in collector current divided by the small change in base current. However, V_{CE} must remain constant when these corresponding changes in current are measured. The current gain (beta) of a common-emitter transistor is expressed simply as a number and is not assigned a specific unit of measurement.

A common-emitter transistor's collector characteristic curves may be used to graphically determine the transistor's current gain (beta). To illustrate this point we will use the curves in Figure 6-4. We will assume that V_{CE} remains constant at 20 volts and that I_B changes from 100 to 125 microamperes. This means that I_C will change from approximately 2.5 milliamperes to approximately 3.6 milliamperes, and the transistor's operating point will move from point A to point B as shown. A total change in I_B of 125-100 or 25 microamperes is therefore accompanied by a total change in I_C from approximately 3.6 mA to 2.5 mA or 1.1 milliamperes. When these values are inserted in the equation for beta we obtain a current gain of:

$$\text{current gain } (\beta) = \frac{1.1 \text{ milliamperes}}{25 \text{ microamperes}}$$

$$= 44$$

Our calculations show that the common-emitter transistor represented by the curves in Figure 6-4 has a current gain (beta) of 44. This is a typical beta value for many small and medium power transistors. However, some transistors may have beta values as low as 10 while others have beta values which exceed 200.

Manufacturers of transistors often specify the beta value for each type of transistor that they produce. The beta values given may be a minimum expected value but average and maximum expected values are sometimes provided. In most cases the manufacturer will also state the transistor's operating conditions along with its beta value since the transistor's beta can change slightly when its operating currents and voltages change.

The beta value just described is a measurement of the transistor's ability to amplify a changing or AC current and (like the common-base transistor) it is often referred to as the transistor's forward current transfer ratio. This AC measurement is also commonly represented by the symbol h_{fe}. However, the transistor's beta can also be expressed as a DC measurement by using fixed (DC) values of I_B and I_C instead of changing (AC) values. In this case the corresponding I_B and I_C values are observed at a specific operating point (possibly point A in Figure 6-4) and their ratio is determined. The DC beta (forward current transfer ratio) is often represented by the symbol H_{FE}.

When the alpha value of a transistor is known but the beta value is not known, it is possible to determine the beta value mathematically instead of plotting it graphically. The following equation may be used to determine beta when alpha is known.

$$\beta = \frac{\alpha}{1 - \alpha}$$

To illustrate the use of this equation we will assume that a transistor has an alpha of 0.98. When this alpha value is inserted in the equation we find that the transistor has a beta of:

$$\beta = \frac{0.98}{1 - 0.98}$$

$$= 49$$

When a transistor's beta is known but its alpha value is not known, the alpha value may be mathematically determined by:

$$\alpha = \frac{\beta}{\beta + 1}$$

To illustrate the use of this equation we will assume that a transistor has a beta of 100. When we insert this beta value into the equation we find that the transistor's alpha is equal to:

$$\alpha = \frac{100}{100 + 1}$$
$$= 0.99$$

When a transistor's beta is known and its alpha must be determined, the mathematical method should be used instead of the graphical method previously described. This is because of the difficulty in obtaining accurate readings from the transistor's characteristic curves.

Beta Cutoff Frequency

Like the common-base transistor previously described, the common-emitter transistor exhibits a decrease in current gain when it is required to amplify AC signals that have sufficiently high frequencies. The frequency required to reduce the common-emitter transistor's current gain (beta) to 70.7 percent of its low frequency value is referred to as the transistor's beta cutoff frequency and is generally represented by the symbol f_{ae}. At frequencies above the f_{ae} value, the gain of the transistor is seriously reduced and its performance is therefore degraded.

It should be pointed out at this time that there are a variety of differences between reference books concerning AC and DC terms. For example, some books use (v) to mean an AC voltage and V to mean DC voltages. Some textbooks refer to AC alpha and beta as h_{fb} and h_{fe} and DC alpha and beta as H_{FB} and H_{FE}. Other reference books use the terms h_{fb} and H_{FB} interchangeably. In the Heathkit/Zenith courses you may also find a mixture of terms. However, the written explanations will clear up any doubts about which parameters we are talking about. In most examples the AC and DC forward current transfer ratios are assumed to be approximately the same and can be used interchangeably.

When any bipolar transistor is alternately connected in a common-base and then a common-emitter arrangement and its re-spective cutoff frequencies (f_{ab} and f_{ae}) are determined, the transistor's f_{ae} will always be much lower than its f_{ab}. In other words, a common-base circuit is capable of amplifying much higher frequencies than a common-emitter circuit. Unfortunately the common-base transistor's current gain (alpha) is much lower than the common-emitter transistor's current gain (beta).

Collector-To-Emitter Leakage Current

The collector characteristic curves in Figure 6-4 show that I_C will decrease to a very low value when I_B is reduced to zero. The small leakage current that flows through the transistor at this time is commonly referred to as the transistor's collector-to-emitter leakage current and is often represented by the symbol I_{CEO}.

Manufacturers often provide the I_{CEO} value of a transistor because it must often be considered when determining how the transistor will operate in certain situations. The I_{CEO} value is determined by using a circuit similar to the one shown in Figure 6-3. However, the base lead of the transistor is left open (disconnected) to insure that I_B is equal to zero.

The leakage current I_{CEO} is comparable to the I_{CBO} value previously described because it is produced by minority carriers and it is temperature sensitive. However, for any given transistor I_{CEO} is much larger than I_{CBO}

and it can have a significant influence on the transistor's operation in certain applications. I_{CEO} will combine with I_C and effectively increase the value of I_C when the transistor is biased for normal operation. It is therefore desirable to select transistors with the lowest possible I_{CEO} values to avoid the problems that excessive leakage can produce.

12. The common-emitter transistor's collector characteristic curves show the relationship between the transistor's collector current (I_C) and collector-to-emitter voltage (V_{CE}) for various values of _____.

13. The collector characteristic curves in Figure 6-4 show that I_C will be slightly less than I_B for any specific value of V_{CE}.

    A. True
    B. False

14. The curves in Figure 6-4 show that a small change in I_B can produce a large change in _____ when V_{CE} remains constant.

15. According to the curves in Figure 6-4, V_{CE} has little control over I_C once the transistor is operating beyond the knee of each curve.

    A. True
    B. False

16. Figure 6-4 shows that within the saturation region, I_C increases rapidly as V_{CE} _____ by only a small amount.

17. The current gain of a common-emitter transistor is often represented by the Greek letter _____.

18. The current gain of a common-emitter transistor is expressed as a ratio and has no unit of measurement.

    A. True
    B. False

19. The mathematical expression $\Delta I_c / \Delta I_B$ is used to represent the current gain of the common-emitter transistor.

    A. True
    B. False

20. Typical common-emitter transistors may have current gains that range from _____ to over _____.

21. A mathematical equation may be used to determine a transistor's beta value when its _____ value is known.

22. The frequency required to reduce the common-emitter transistors current gain to 70.7% of its low frequency value is referred to as the transistor's _____ _____ _____.

23. A transistor's collector-to-emitter leakage current is represented by the symbol _____.

Characteristics Of Common-Collector Circuits

When a bipolar transistor is connected in a common-collector (also called emitter-follower) arrangement, the input signal voltage is applied between the transistor's base and collector and effectively varies the transistor's emitter and collector currents but it is the emitter current that is used as the output current. The emitter current flows through a load resistance and the voltage developed across this resistance serves as the output voltage. This output voltage effectively appears between the transistor's emitter and collector. The common-collector arrangement cannot produce an increase in signal voltage because the emitter voltage tends to track or follow the base voltage as explained in a previous unit, however, the circuit arrangement can produce a substantial increase in current and power.

The common-collector arrangement is not used as often as the common-emitter arrangement because it cannot provide voltage amplification. However, the arrangement is used in certain applications where a signal source with a high internal resistance must supply power to a low resistance load. The common-collector circuit with its inherently high input resistance and its low output resistance serves as a buffer and effectively prevents the load from drawing too much current from the source but at the same time allows approximately the same source (signal) voltage to be applied to the load. Since the common-collector circuit allows the same signal voltage supplied by a high resistance source to appear across a low resistance load, the circuit must provide a substantial increase in signal current.

When the common-collector arrangement is compared with the common-emitter circuit, it is possible to mathematically derive an equation which shows how to determine the current gain of the common-collector transistor. We previously discovered that the common-emitter transistor's current gain (beta) is determined by dividing a change in output (collector) current by a corresponding change in input (base) current and is expressed mathematically as:

$$\text{current gain } \beta = \frac{\Delta I_C}{\Delta I_B}$$

However, the current gain of a common-collector transistor must be equal to the change in output (emitter) current divided by a corresponding change in input (base) current and therefore must be represented mathematically as:

$$\text{current gain} = \frac{\Delta I_E}{\Delta I_B}$$

Since a transistor's emitter current must be equal to the sum of its base and collector currents we can express the emitter current as:

$$I_E = I_B + I_C$$

Therefore it is possible to express the current gain of a common-collector amplifier as:

$$\text{current gain} = \frac{\Delta (I_B + I_C)}{\Delta I_B}$$

which can also be expressed as:

$$\text{current gain} = \frac{\Delta I_B + \Delta I_C}{\Delta I_B}$$

This last equation can be further reduced by dividing the term (ΔI_B) to obtain the final equation for current gain which is expressed as:

$$\text{current gain} = 1 + \frac{\Delta I_C}{\Delta I_B}$$

or

$$\text{current gain} = 1 + \beta$$

Therefore, the current gain of a common-collector transistor is equal to 1 plus the transistor's beta. For all practical purposes when the transistor's beta is high (over 30) the gain of the transistor when it is connected in a common-collector arrangement will be equal to its beta value. Therefore common-collector and common-emitter transistors have approximately the same current gain.

The equation that was just derived is for the AC current gain of the transistor since the transistor's AC beta ($\Delta I_C / \Delta I_B$) is used. However, the transistor's DC beta (I_C / I_B) may also be used in the common collector configuration.

Manufacturers of transistors seldom provide information relating to all three circuit configurations. In most cases, the manufacturer specifies the features or characteristics of only one circuit arrangement (usually the common-emitter). Therefore it is often necessary to mathematically determine the characteristics of one configuration from information provided on another configuration.

The calculations just given show how this is done in the case of current gain. The equation obtained is often referred to as a conversion formula since it shows how to convert from common-emitter gain to common-collector gain. The previous equations showed how to determine alpha when beta is known or beta when alpha is known and are also conversion formulas.

Input Resistance

Although the common-collector circuit exhibits a substantial current gain, it is most often used because it provides a high input resistance and a low output resistance. The approximate input resistance of a common-collector circuit can be obtained by using the following equation.

$$R_{in} = \beta R_L$$

This equation states the input resistance (R_{in}) is approximately equal to the transistor's beta times the value of the load resistance connected in the transistor's emitter (output) circuit. This load resistance is represented by the designation R_L.

A basic common-collector circuit is shown in Figure 6-7. This circuit is drawn in a slightly different manner than the common-collector circuit shown in the previous unit; however, it is basically the same circuit. The transistor is biased by voltages V_{BB} and V_{CC}, and a load resistance (R_L) has been connected between the transistor's emitter and collector. This resistor is electrically between the emitter and collector because V_{CC} is effectively shorted as far as input signals are concerned. The transistor's collector is therefore common to the circuit's input and output. A signal source and its internal resistance (R_i) is also connected to the input of the circuit as shown.

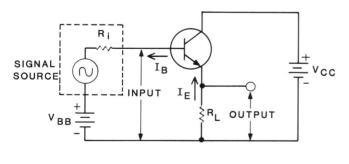

Figure 6-7 A basic common-collector circuit.

We will assume that the transistor in Figure 6-7 has a beta of 50 and that R_L is equal to 2000 ohms. When you substitute these values into the equation given above you obtain an input resistance of approximately:

$$R_{in} = (50)(2000) = 100,000 \text{ ohms}$$

An input resistance of 100,000 ohms (as determined above) is typical for a common-collector transistor. Although this value may seem high, it can be made much higher by using a transistor with a higher beta or by using a higher value of load resistance.

The value of input resistance (100,000 ohms) just calculated represents only the input resistance of the transistor with the specified value of R_L. The input resistance of the circuit will be changed by a substantial amount if external resistors (such as those used to set the transistor's forward bias) were connected in the circuit. In other words, the input resistance of the entire circuit will be different (usually lower) than the input resistance of the transistor.

Output Resistance

It is much more difficult to calculate the common-collector transistor's output resistance (the resistance seen when looking back into the transistor's output terminals). It is necessary to consider the transistor's gain, its input resistance, and even the internal resistance (R_i) of the signal source that is connected to the input of the transistor. Furthermore, this calculation is usually made with R_L removed from the circuit. Then the value of output resistance obtained is considered to be in parallel with R_L and this parallel combination is used to determine the output resistance of the circuit. The two most important factors which control the output resistance of a common-collector circuit are the internal resistance (R_i) of the signal source and the transistor's beta. See Figure 6-7. A rough estimate of the output resistance can be obtained by simply dividing R_i by the transistor's beta as shown:

$$R_{out} = \frac{R_i}{\beta}$$

This equation provides reasonably accurate results as long as R_i is quite large. Once this equation is used to determine R_{out}, the parallel resistance of R_{out} and R_L must be calculated to determine the overall output resistance of the circuit.

The value of R_{out} is usually quite low (often less than 100 ohms) in most common-collector circuits. Therefore the overall output resistance of the circuit (R_{out} and R_L in parallel) is generally much lower than the value of R_L. Since the simplified equation shows R_{out} to be equal to $R_i/(\beta)$, it follows that the source resistance (R_i) is transformed to a lower value by a factor equal to the transistor's beta. However, both the input resistance (described earlier) and the output resistance are affected by changes in signal frequency because the transistor's beta is likewise affected by frequency changes. In general, when the signal frequency exceeds the f_{ae} value, the transistor's beta drops rapidly, thus, causing R_{in} to decrease and R_{out} to increase. The common-collector arrangement therefore provides a high input resistance and a low output resistance over a frequency range which extends up to f_{ae}.

Maximum Transistor Ratings

To this point we have examined some of the important characteristics of each circuit configuration (common-emitter, and common-collector). However, we assumed that each transistor was working within safe operating limits. Unfortunately, transistors can be damaged if they are subjected to excessively high currents or voltages and it is important to know just how much current or voltage each device can withstand. Manufacturers usually specify the maximum safe operating currents and voltages for each device and may also specify various power and temperature limitations. These maximum safe values are often referred to as maximum transistor ratings. Some of the most important ratings are described in the following paragraphs.

Collector Breakdown Voltage

The amount of collector-to-base reverse bias voltage required to produce a sharp increase in collector-to-base current is referred to as the transistor's collector breakdown voltage.

The sudden increase in current occurs because the transistor's collector junction breaks down in the same way that an ordinary semiconductor diode breaks down when it is subjected to a sufficiently high reverse bias voltage. The collector breakdown voltage is measured with the transistor's emitter open so that I_E is equal to zero and it is usually specified at some value of reverse leakage current. The collector breakdown voltage is often designated as V_{CBO} or βV_{CBO}.

Emitter Breakdown Voltage

The amount of reverse bias voltage required to break down the emitter-to-base junction is referred to as the transistor's emitter breakdown voltage. The emitter breakdown voltage is measured with the transistor's collector open so that I_C is equal to zero and it is often represented by the symbol V_{EBO} or the symbol βV_{EBO}. Furthermore, V_{EBO} is usually specified at some value of reverse leakage current.

The V_{EBO} value specified by the manufacturer should never be exceeded (even momentarily) or the transistor could possibly be damaged. This same consideration is also true for the V_{CBO} rating previously described. In general, transistors should always be subjected to external voltages which are well below these maximum breakdown ratings.

Maximum Collector and Emitter Currents

Manufacturers usually specify the maximum value of collector current (I_C) and emitter current (I_E) that a transistor can safely handle. If these current values are exceeded, the transistor could be permanently damaged. The transistor's operating currents should be well below these maximum values.

Maximum Collector Dissipation

A bipolar transistor dissipates power in the form of heat because it is required to conduct current while being subjected to external voltages. Practically all of the dissipation occurs at the transistor's reverse-biased collector junction and can be easily calculated by multiplying the transistor's collector-emitter voltage by the collector current flowing through the device.

The amount of power that a transistor can safely dissipate at its collector junction is often referred to as the transistor's maximum collector dissipation rating. Typical power ratings for transistors will range from several hundred milliwatts to more than 100 watts at an operating temperature of 25 degrees centigrade. When the operating temperature exceeds 25 degrees centigrade, the power rating of the transistor must be derated. In other words it is necessary to calculate the power rating of the device (which will be lower) for the higher operating temperatures. This can usually be accomplished by using information supplied by the manufacturer of the device in the form of a component data or specification listing.

Temperature Ratings

Bipolar transistors operate within certain temperature ranges. When the temperature of a transistor is too high or too low, the device may not operate efficiently, and in fact, could be damaged. In general, transistors made of silicon materials are capable of operating over a wider temperature range than germanium transistors. It is also common practice for manufacturers to specify the safe temperature range of each device. In fact, manufacturers usually indicate the temperature range in which a transistor should be operated or stored.

Manufacturers are usually specific when indicating a transistor's operating temperature range. In most cases, the transistor's permissible junction or case temperatures are specified. For example, a typical silicon transistor may have an operating (junction) temperature range of 65 degrees to 200 degrees centigrade. In situations where the transistor's operating temperature tends to approach or exceed the upper end of this range, the transistor is usually attached to a suitable heat sink. The heat sink increases the transistor's ability to radiate heat and therefore helps to cool the device. The temperature affects the power dissipation of the transistor as well as the other characteristics such as beta. Beta is usually proportional to temperature.

Current Gain – Bandwidth Product

We indicated earlier that the frequency response of a transistor is designated by its alpha or beta cut off frequency. The transistor will provide suitable gain when it is operated below these frequencies. Another way of expressing the frequency response of a transistor is the specification f_T, the current gain-bandwidth product. This specification is more often seen on manufacturer's data sheets than the alpha or beta cut off frequencies. The term f_T simply indicates the frequency where the current gain in the common emitter mode is one. This is the maximum operating frequency of the transistor. The current gain - bandwidth product is basically constant, therefore as the operating frequency decreases, the current gain increases by the amount necessary to keep f_T constant.

24. The current gain of a common-collector transistor is equal to 1 plus the transistor's _____.

25. If a transistor has a high beta, the current gain of the device will be approximately equal to its beta value when it is connected in a common-collector arrangement.

    A. True
    B. False

26. Manufacturers of transistors usually specify the important characteristics for each of the three basic circuit configurations.

    A. True
    B. False

27. The input resistance of a common-collector transistor is approximately equal to the transistor's beta times the _____ connected in its emitter circuit.

28. The approximate output resistance of a common-collector transistor may be determined by dividing the _____ of the signal source connected to the transistor by the transistor's beta.

29. The symbol β V_{CBO} is used to indicate a transistor's _____ _____ voltage.

30. The symbol β V_{EBO} is used to indicate a transistor's _____ _____ voltage.

31. Most of the power dissipated within a transistor occurs at the transistor's _____ junction.

32. The amount of power that a transistor can safely dissipate is referred to as the transistor's maximum _____ _____ rating.

33. A transistor's operating temperature affects its _____ rating and _____.

34. A transistor's upper frequency limit is indicated by the term _____.

Desk-Top Experiment 4
Converting A Pictorial Diagram
To A Schematic Diagram

Introduction

This exercise will help you convert between pictorial wiring diagrams and schematic diagrams. As you know, a schematic diagram is an electrical representation of a physical circuit.

It is not always obvious what the circuit arrangement is when viewed in its pictorial or actual form. It is also not always obvious which points are the inputs and outputs of the circuit.

Objectives

1. To convert the pictorial circuit shown in Figure 6-8 into its schematic representation.

2. To identify the circuit arrangement and label the points of interest.

3. To explain the advantages of the pictorial diagram over a schematic diagram.

Procedure

Convert the pictorial diagram shown in Figure 6-8 into a schematic diagram.

1. Pick a starting point. One possibility is to draw the transistor's symbol in the center of the space allotted for your schematic drawing.

2. Label the 3 leads as emitter, collector, and base.

3. What type of transistor is it (NPN or PNP)? Note, that on the pictorial diagram the arrow on the emitter is not shown. However, enough information is given to determine if the transistor is a PNP or an NPN type.

4. Draw a schematic representation of the collector circuit's components and current path.

5. Draw a schematic representation of the base circuit's components and circuit path.

6. Draw a schematic representation of the emitter circuit's components and circuit path.

Discussion

The reason for starting with the transistor in the center is that this normally splits the circuit approximately in half. This usually ensures that you will have room to draw the rest of your schematic diagram.

Labeling the emitter, collector, and base leads helps by dividing the circuit into current paths. This makes it easier to draw the individual current paths. Each current path will end at either ground or a reference voltage.

If the collector's current path is tied to ground, the circuit is in the form of a common collector arrangement. If the collector's current path ends at the positive voltage source, the transistor must be an NPN type. If the current path leads to a negative source voltage the transistor must be a PNP type. Once the type of transistor is known, you should know where to put the arrow and which way it should point.

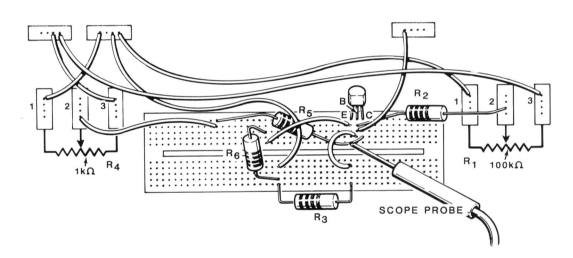

Figure 6-8 Pictorial diagram for Desk-Top Experiment 4.

Procedure (Cont.)

7. Where 2 or more components or current paths are connected to the same point, draw a small circle enclosing the tie point and black in the circle.

8. Label the components as to R_1, R_2, etc.

9. Examine Figure 6-8 carefully to determine if there are any components that are polarity sensitive. Mark your schematic drawing accordingly.

10. Combine common points together.

Discussion

The blacked in tie point is a current node. A current node is a point in the circuit where a current path splits into two or more parallel paths.

Where 2 lines cross it should be obvious on your drawing, if the lines are connected at the crossover point, or if it is simply a crossover in route to another connection.

The letter used in component labeling indicates the type of component, R for resistors, C for capacitors, L for inductors, and so forth. The number designator R_1 can be used to keep track of how many of the same type components are used in the circuit. It is also a valuable tool to distinguish between resistors in discussions.

Next, a point should have been picked that is close to all of the points that are to be connected to the same voltage source. They should all be joined to a single point and then a single line connected from the common point to the source. Additional lines that go to the same place are unnecessary and only serve to clutter your schematic drawing. The same is true for all of the points tied to ground. Examine Figure 6-9 and Figure 6-10. It is obvious in Figure 6-10 that there is only one voltage source. In Figure 6-9 it is not obvious that the same voltage can be used for base biasing and the collector's supply voltage.

Procedure (Cont.)

Evaluate your schematic drawing as to:

11. Type of circuit arrangement.

12. Function of R1.

13. Function of R2 and R3.

14. Function of R4.

15. Function of R5.

16. Function of R6.

17. The phase of the output waveform.

18. Assume the output is taken across R6, the output waveform would be?

Discussion

The circuit is a common emitter arrangement. Since the input is applied to the base and the output is taken from the collector, the emitter is common to both the input and output.

R_2 and R_3 form a voltage divider that is connected to a variable supply voltage and is used to set the static bias voltage. The static bias in turn sets the output reference voltage.

R_4 provides a variable collector voltage source.

R_5 is considered to be the fixed collector load resistor, but it actually combines with the resistance between terminals 2 and 3 of the potentiometer R_4 to provide a resistance that could be varied. If terminal 1 of R_4 was left open instead of connected to ground, the load resistor could be increased by as much as 1 k ohm.

R_6 provides self bias and helps stabilize the circuit by reducing current flow in the circuit. As the input voltage increases (more positive) current through the transistor increases. Since all of the current flowing through a transistor flows through the emitter lead, the voltage drop across R_6 increases (becomes more positive). This has the overall effect of holding the increasing current to a lower value, but it increases at an almost linear rate.

Since it is a common emitter arrangement, the output would be shifted by 180 degrees when compared to the input waveform. The amplitude of the output would also be increased. The amount of increased amplitude would depend on the circuit's components and the transistor's beta.

If the output was taken across R_6, the voltage waveform would have the same phase as the input waveform and V_o would be slightly less than V_{in}.

Summary

Examine the circuit in Figure 6-9. Your schematic drawing should resemble this schematic diagram of the pictorial circuit shown in Figure 6-8. An important point to remember is that a schematic diagram is an equivalent electrical description of an actual circuit and has little resemblance to the physical layout of the actual circuit. With a little practice you will be able to wire a circuit from either a pictorial or schematic wiring diagram. The pictorial diagram is useful in situations where workers, who are unfamiliar with schematic symbols and circuit functions, assemble electronic products. If a product is extremely complex, it can be drawn as a series of smaller pictorials.

The transistor should be in the center of your drawing because this is where the circuit action and evaluation will take place. You should have noticed that it is important to mark the emitter, base, and collector leads. This helps you to see the input and output points of your circuit, and it also shows you what the circuit arrangement is. Your circuit is a common emitter arrangement using an NPN transistor.

The node points in the schematic are the locations where the current takes more than one current path. This shows the parallel combinations that your circuit contains.

In your circuit there is only one voltage source. In your circuit the transistor is the only polarity sensitive component. However, in many circuits there will be polarity sensitive components such as electrolytic and tantalum capacitors and diodes. Care must be exercised to ensure that polarity sensitive components are properly wired into circuits. When improperly installed (reversed) they may overheat, and in the case of some components they may explode. Besides the components themselves being damaged it is probable that they will cause damage to other circuit components. It is even possible that they could explode and cause serious injury to you or anyone that is nearby.

Notice in Figure 6-10 that the multiple grounds have been combined to reduce the number of wires in the drawing. This is also true of the source wiring. In this example it did not simplify the circuit, but it did establish that only one voltage supply source is required.

The circuit base bias can be varied by adjusting the value of R1. Increasing the voltage available at pin 2 of potentiometer R1 would increase the transistor's rate of conduction.

R_4 can be used to adjust the supply voltage that is applied to the transistor. Care should be taken so that clockwise rotation causes the output voltage to increase. The direction of rotation can be reversed by reversing the wires connected to pins 1 and 3 of potentiometer R_4.

R_5 is the fixed load resistor for the transistor's collector.

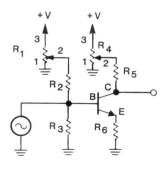

Figure 6-9 Schematic diagram for the circuit shown in Figure 6-8.

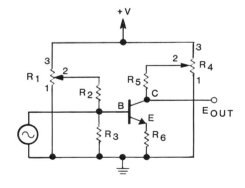

Figure 6-10 Final schematic diagram for Figure 6-8.

R_6 is used to self bias the transistor's emitter-base junction. This is degenerative feedback and it reduces the gain of the amplifier by reducing the current flowing through the transistor. This provides thermal stability because it reduces the change in current flow through the transistor. This causes the transistor to operate at a somewhat controlled temperature and results in reducing the transistor's operating temperature. The base bias voltage is the difference between the voltage dropped across R_3 and the voltage dropped across R_6.

A bipolar transistor may be connected in one of three basic circuit arrangements/configurations. The basic configurations are referred to as the common-base circuit, the common-emitter circuit, and the common-collector circuit. The common-base and common-emitter arrangements are used to provide voltage amplification of input signals, and the common-collector arrangement is often used to match a high resistance source to a low resistance load.

The amount of current amplification or gain that a transistor can provide can be expressed as a ratio of its input to output currents. In the case of a common-base transistor this ratio is referred to as the transistor's alpha (α) and it is always less than 1. In the case of a common-emitter transistor, the ratio is known as the transistor's beta (β). Furthermore, the alpha and beta values for a transistor can be determined by using AC (changing) values of current or DC (fixed) values of current.

A transistor's current gain tapers off when the device is subjected to a sufficiently high signal frequency. The frequency at which a common-base transistor's current gain (α) drops to 70.7% of its low frequency value is referred to as the transistor's alpha cutoff frequency. The frequency at which a common-emitter transistor's current gain (beta) drops to 70.7% of its low frequency value is referred to as the transistor's beta cutoff frequency.

Leakage current is undesirable in a transistor but it exists in both NPN and PNP transistors that are made from either silicon or germanium. In general, the leakage current is higher in germanium transistors than it is in silicon transistors and in each case the leakage current increases with temperature. The two most important leakage currents that occur in a transistor are represented by the symbols I_{CBO} and I_{CEO}. I_{CBO} is the transistor's collector-to-base leakage current when the transistor's emitter is open. I_{CEO} is the transistor's collector-to-emitter leakage current when the transistor's base is open.

The high input resistance and low output resistance of the common-collector transistor makes the circuit suitable for use in matching resistances. The input resistance of a common-collector transistor is approximately equal to the transistor's beta times the value of the load resistance (R_L) that is connected to the transistor's emitter. The common-collector transistor's output resistance is approximately equal to the internal resistance of the signal source (R_i) that is connected to the transistor's input leads divided by the transistor's beta.

The most important data on a manufacturer's transistor data sheet is its maximum current, voltage, power, temperature, and frequency ratings.

Unit 7
Field Effect Transistors

Contents

Introduction

In this unit you will examine another important solid-state component that is commonly referred to as a field effect transistor. The field effect transistor (commonly called an FET pronounced fet) operates on a principle that is completely different from that of the conventional bipolar transistor. The FET is a three-terminal device that is capable of providing amplification and it can compete with conventional bipolar transistors in many applications. A basic understanding of FET operation and construction is therefore essential if you are to have a well rounded background in solid-state components.

Basically there are two types of field effect transistors. One type is known as a junction field effect transistor but is commonly referred to as a junction FET or simply a JFET. The second type is known as an insulated gate field effect transistor (IGFET), although it is frequently referred to as a metal-oxide semiconductor field effect transistor (MOSFET). As you proceed through this unit you will examine both of these basic FET devices. You will learn how each device is constructed and how it operates. You will also demonstrate the operation of a typical JFET and a typical IGFET.

When you have completed this unit on field effect transistors you should be able to:

1. Describe how a junction FET operates.

2. Use a FET's drain characteristic curves to determine the transconductance of the device.

3. Properly bias both N-channel and P-channel JFET's.

4. Explain the meaning of the expressions V_{GS} (off and V_p).

5. Explain the basic difference between the JFET and MOSFET.

6. Describe the difference between depletion- and enhancement-mode IGFETs.

7. Properly bias both depletion-mode and enhancement-mode IGFETs.

8. Name the three basic FET circuit arrangements.

9. Explain the advantages and disadvantages of FETs when compared to bipolar transistors.

10. Build a circuit using an FET as the amplifing component.

11. From a schematic point out the:

    N-channel FET
    P-channel FET
    depletion-mode IGFET
    enhancement-mode IGFET

The junction FET (also called a JFET) finds many applications in electronic circuits. This device is constructed from N-type and P-type semiconductor materials and like the conventional bipolar transistor, it is capable of amplifying electronic signals. However, the junction FET is constructed in a different manner than a bipolar transistor and the device operates on an entirely different principle.

A basic understanding of junction FET construction is necessary in order to understand how the device operates. Therefore, we will first consider the physical aspects of the junction FET and then we will consider its electrical characteristics.

Construction of a JFET

The construction of a junction FET begins with a lightly doped semiconductor material (usually silicon) which is referred to as the substrate. The substrate simply serves as a platform on which the remaining electrodes are formed and the substrate can be either a P-type or an N-type material. Through the use of diffusion and epitaxial growth techniques, an oppositely doped region is formed within the substrate to create what is effectively a PN junction. However it is the unique shape of this PN junction that is important.

The structure that is created by the process described above is shown in Figure 7-1. The region that is embedded in the substrate ma-

terial is U-shaped so that it is flush with the upper surface of the substrate at only two points.

The embedded region actually forms a channel of oppositely doped semiconductor material through the substrate. Therefore, the embedded region is generally referred to as simply a channel as indicated in Figure 7-1. When this channel is made from an N-type material and embedded in a P-type substrate, the entire structure is known as an N-channel junction FET. However; when a P-type channel is implanted in an N-type substrate, the device becomes a P-channel junction FET. Therefore, a junction FET is similar to a conventional bipolar transistor in that it can be constructed in two different ways. The junction FET is either an N-channel or a P-channel device while the bipolar transistor is either an NPN or a PNP device.

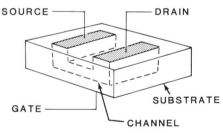

Figure 7-1 Basic construction of a Junction FET.

The construction of a basic junction FET is completed by making three electrical connections to the device as shown in Figure 7-1. One wire or lead is attached to the substrate as shown. This connection is referred to as the gate. Then wires are attached to each end of the channel as shown. These two leads are referred to as the source and the drain. In most junction FET's the channel is geometrically symmetrical and it makes no difference which end of the channel is used as the source and which is used as the drain. Only in special types of FET's will the channel be asymmetrical and when using these special devices the source and drain leads cannot be interchanged.

Operation of a JFET

Like a conventional bipolar transistor, a junction FET requires two external bias voltages for proper operation. One voltage source is normally connected between the source and drain leads so that a current is forced to flow through the channel within the device. The second voltage is applied between the gate and the source and it is used to control the amount of current flowing through the channel.

Refer to Figure 7-2A. This figure shows a cross section view of an N-channel junction FET and its required operating voltages. Notice that an external voltage source is connected between the drain (D) and the source (S) leads and this drain-to-source voltage is represented by the symbol V_{DS}. The voltage V_{DS} is connected so that the

source is made negative with respect to the drain. This voltage causes a current to flow through the N-type channel because of the majority carriers (free electrons) within the N-type material. This source-to-drain current is commonly referred to as the FET's drain current and is represented by the symbol I_D. The channel simply appears as a resistance to the supply voltage V_{DS}.

Also notice in Figure 7-2A that a voltage is applied between the gate (G) and the source (S) of the FET. This gate-to-source voltage (designated as V_{GS}) causes the P-type gate to be negative with respect to the N-type source. Since the source is effectively just one end of the N-type channel, V_{GS} effectively reverse biases the PN junction formed by the P-type gate and the N-type channel. This reverse bias voltage causes a depletion region (an area devoid of majority carriers) to form within the vicinity of the PN junction. As shown in Figure 7-2A this depletion region spreads inward along the length of the channel. Although two depletion regions appear to exist, only one is created. This depletion region extends around the wall of the N-type channel since all sides of the channel are in contact with the P-type substrate (which serves as the gate). Furthermore, the depletion region will be somewhat wider at the drain end of the channel than at the source end. This is because V_{DS} effectively adds to V_{GS} so that the voltage across the drain end of the PN junction is higher than the voltage across the source end of the junction.

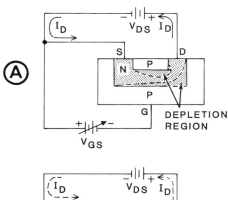

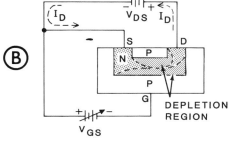

Figure 7-2 A properly biased N-channel Junction FET.

The size of the depletion region in Figure 7-2A is controlled by voltage V_{GS}. When V_{GS} increases, the depletion region increases in size. When V_{GS} decreases, the depletion region decreases in size. Furthermore, when the depletion region increases in size, the N-type channel is effectively reduced in size (fewer free electrons are available) and less current will be able to flow through the channel. The opposite is true when the depletion region decreases in size. This means that V_{GS} may be used to effectively control the drain current (I_D) flowing through the channel. An increase in V_{GS} results in a de-

crease in I_D and vice-versa. We can say that V_{GS} controls the resistance of the channel.

Remember that V_{GS} reverse biases the PN junction formed by the gate and the channel. Therefore, only an extremely small (almost insignificant) leakage current flows from the gate to the source.

It is important to note that a voltage (V_{GS}) is used to control the drain current (I_D) in a junction FET. In normal operation the voltage applied between the gate and the source (V_{GS}) serves as an input voltage, to control the device. The drain current (I_D) represents the output current which can be made to flow through a load. This action is considerably different from the action that takes place in a bipolar transistor. In the bipolar transistor, an input current (not a voltage) is used to control an output current. Also; since the gate-to-source junction of the FET is reverse-biased by V_{GS}, the FET has an extremely high input resistance. This is just the opposite of a bipolar transistor which has a forward-biased emitter-to-base junction and therefore a relatively low input resistance. The gate-to-source voltage (V_{GS}) must never be reversed so that the PN junction formed by the gate and the channel becomes forward-biased. This would result in a relatively large current through the junction which would cause the input resistance of the device to drop to a low value and the gain of the device would also be significantly reduced.

Gate-to-Source Cutoff Voltage

As mentioned earlier, when V_{GS} is increased, the depletion region within the FET increases in size and allows less drain current (I_D) to flow. If V_{GS} is increased to a sufficiently high value, the depletion region is increased in size until the entire channel is depleted of majority carriers as shown in Figure 7-2B. This causes I_D to decrease to an extremely small value and for all practical purposes is reduced to zero. The gate-to-source voltage required to reduce I_D to zero (regardless of the value of V_{DS}) is referred to as the gate-to-source cutoff voltage and is represented by the symbol $V_{GS(off)}$. Manufacturers of FET's usually specify the applicable cutoff voltage for each type of FET that they produce.

Pinch-Off Voltage

The drain-to-source voltage (V_{DS}) also has a certain amount of control over the depletion region within the junction FET. The effect of V_{DS} can be noted in Figures 7-2A and 7-2B since the depletion region is wider near the drain than it is near the source. As explained earlier, this occurs because V_{DS} is in series with V_{GS} thus causing a greater voltage to exist across the PN junction near the drain. If V_{DS} increases in value, the action becomes even more pronounced and can affect the drain current (I_D) flowing through the device.

If V_{DS} is increased from zero to higher values, I_D is also increased. However, a continued increase in V_{DS} will not result in a constant rise in I_D. Instead, a point is soon reached where I_D levels off and then increases only slightly as V_{DS} continues to rise. This action occurs because the size of the depletion region increases (especially near the drain) as V_{DS} becomes so depleted of majority carriers that it will not allow I_D to increase proportionally with V_{DS}. In other words a point is reached where the resistance of the channel effectively begins to increase as V_{DS} increases thus causing I_D to increase at a much slower rate.

Since I_D levels off because the depletion region expands and effectively reduces the channel width, I_D is said to be pinched-off. The value of V_{DS} required to pinch off or limit I_D is referred to as the pinch-off voltage and is represented by the symbol V_P. Manufacturers usually provide the value of V_P for a given FET for a gate-to-source voltage (V_{GS}) of zero. V_P is measured by shorting the gate and source leads. This means that I_D will increase up to its maximum possible value (when, $V_{GS} = 0$) and then level off. When V_{GS} is equal to zero, the drain current flowing through the FET is often identified as I_{DSS} instead of I_D. Furthermore, manufacturers often provide the FET's I_{DSS} value when V_{DS} is equal to or greater than V_P. In this case I_{DSS} represents the FET's maximum drain current when V_{GS} is equal to zero.

In practice, the value of V_P (when $V_{GS} = 0$) will always be close to the value of $V_{GS(off)}$

for any given FET. In fact these two quantities may be interchanged in any calculation in which either quantity is involved. This means that when V_{GS} is equal to or greater than V_P; the drain current (I_D) will be effectively reduced to zero. Also, when V_P is equal to the $V_{GS(off)}$ value, the drain-current (I_D) flowing through the device will be effectively pinched off.

Drain Characteristic Curves

In the previous unit you saw how a set of collector characteristic curves could be used to show the relationship between the input and output currents and the output voltage associated with a bipolar transistor. A similar set of curves can be used to show the relationship between V_{GS}, V_{DS}, and I_D in a junction FET. In this case they are referred to as drain characteristic curves.

A typical set (sometimes called a family) of drain characteristic curves is shown in Figure 7-3. These curves show how I_D and V_{DS} vary in relation to each other for various values of V_{GS}. Each curve is formed by alternately adjusting V_{GS} to a specific value and then increasing V_{DS} from zero to some maximum value while observing the change in I_D. Notice that when V_{GS} is equal to zero, I_D increases rapidly as V_{DS} increases from zero. However, I_D soon levels off at some maximum value as shown. When this point is reached, the corresponding I_{DSS} and V_P values previously described are obtained.

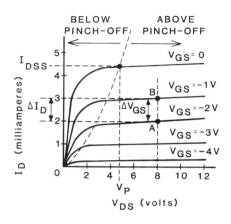

Figure 7-3 Typical drain characteristic curves for an N-channel FET.

The remaining curves in Figures 7-3 are plotted for higher values of V_{GS}. Notice that for each higher value of V_{GS}, I_D levels off at a lower value so that the corresponding pinch-off voltage (V_P) must also be lower. The dashed line that curves upward and to the right crosses each curve at the approximate point where I_D is pinched off. When the FET is biased so that it is operating to the left of this dashed line, the device is said to be operating below pinch-off. This region to the left of the dashed line is sometimes referred to as the ohmic region or the triode region.

When the FET is biased to operate to the right of the dashed line, the device is operating above pinch-off. This region to the right of the line is often referred to as the pinch-off region. In most applications the FET is biased so that it operates above pinch-off or in other words within the pinch-off region. Operation within the pinch-off region is assured by simply making V_{DS} higher than V_P or $V_{GS(off)}$.

As shown in Figure 7-3, the drain current (I_D) will be maximum (for any specific value of V_{DS}) when V_{GS} is equal to zero. However, I_D will decrease as V_{GS} increases in value. As explained earlier, this action results because the channel within the FET becomes depleted of majority carriers as V_{GS} increases, thus offering a higher resistance to the flow of current. Since the operation of the junction FET is controlled by varying the depletion region within the device, the junction FET is said to operate in the depletion mode.

Transconductance

Like the bipolar transistors that were discussed previously, the FET is most often used to amplify electronic signals. In the case of a bipolar transistor, this amplifying ability can be expressed mathematically as a ratio of input and output currents (the transistor's alpha or beta). In the case of the FET, a similar mathematical relationship can also be used.

The amplifying ability of an FET is measured by noting the effect that the gate-to-source voltage (V_{GS}) has on drain current (I_D). V_{GS} is varied a small amount and the corresponding change in I_D is observed. Then these two quantities are expressed as a mathematical ratio. This ratio is com-

monly referred to as the FET's transconductance and is expressed mathematically as:

$$gm = \frac{\Delta I_D}{\Delta V_{GS}}$$

This equation simply states that the transconductance (designated as gm) is equal to a small change in I_D divided by a corresponding change in V_{GS}. Although not shown in this equation, the FET's drain-to-source voltage (V_{DS}) must be held constant when these changes are observed. Furthermore, the transconductance is expressed in units called mhos. The term mho is the reciprocal of ohms or 1/ohms is expressed in mhos.

Although the quantities I_D and V_{GS} could be measured in a test circuit, these quantities can also be determined graphically by referring to an applicable set of drain characteristic curves. To demonstrate how this is accomplished we will use the typical curves shown in Figure 7-3. We will determine the transconductance of the FET within the pinch-off region (above pinch-off) since this is the region most commonly used.

We will assume that V_{DS} remains constant at 8 volts and that V_{GS} changes from 1 volt to 2 volts as indicated at points A and B in Figure 7-3. This change in V_{GS} causes I_D to change from approximately 3 to 2 milliamperes. A total change in V_{GS} of 2 − 1 or 1 volt corresponds to a change in I_D of 3 − 2 or 1 milliampere. When these corresponding changes in I_D and V_{GS} are inserted in the equation given above, we obtain a transconductance of:

$$gm = \frac{1\ mA}{1\ V}$$

$$= \frac{.001}{1}$$

$$= .001\ mho\ or\ 1000\ micromhos$$

The transconductance is therefore equal to 1000 micromhos between points A and B, however, this value will vary slightly at different operating points on the curves. This is because the curves are not equally spaced within the pinch-off region.

The greater the change in I_D for a change in V_{GS}, the higher the gain of the FET. In general a high gm is a desirable FET characteristic.

Symbols

To this point we have examined only the N-channel junction FET. The operation of a P-channel FET has not been discussed because it operates in the same manner as the N-channel device and it has the same basic characteristics. The primary difference is in the manner in which the drain current (I_D) flows through the channel. In a P-channel FET, I_D is supported by the movement of holes in a P-type channel. However, these holes are still the majority carriers within the P-type channel just as the electrons are the majority carriers in the N-channel FET. Also, the P-channel FET has a gate (or sub-

strate) that is formed from N-type material. This of course is opposite to the conditions that exist within an N-channel FET. This means that the polarities of the bias voltages (V_{GS} and V_{DS}) are exactly opposite for the N-channel and P-channel devices.

An N-channel junction FET and a P-channel junction FET are shown in Figure 7-4 as they would appear in a circuit diagram or schematic. The required bias voltages for each device are also shown. Notice that the symbol used for each device is almost identical. The only difference is the direction of the arrow on the gate (G) lead. The N-channel junction FET symbol shown in Figure 7-4A uses an arrow that points inward. However, the P-channel junction FET symbol in Figure 7-4B uses an arrow that points outward.

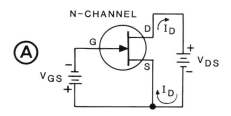

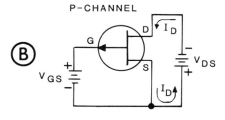

Figure 7-4 Schematic representation of properly biased N-channel and P-channel junction FET's.

Also notice that the polarities of the bias voltages are exactly opposite. The N-channel FET must be biased so that its drain (D) is positive with respect to its source (S) and its gate (G) must be negative with respect to the source. This negative potential on the gate accounts for the use of the minus signs before the V_{GS} values in Figure 7-3.

The P-channel FET must be biased so that its drain is negative with respect to its source and its gate must be positive with respect to the source. The drain current (I_D) will therefore flow in a direction that is opposite to the drain in an N-channel FET. However, the majority carriers (holes) within the P-type channel always move from the source to the drain just like the majority carriers (electrons) in the N-channel device.

1. Junction field effect transistors are usually made from germanium.

   A. True
   B. False

2. The construction of a junction FET usually starts with a section of lightly doped N-type or P-type material known as a _____.

3. In a junction FET, current is made to flow through a _____ that is implanted within the device.

4. The three leads associated with the junction FET are called the _____, _____, and _____.

5. In most junction FET's, the _____ and _____ leads may be interchanged.

6. An N-channel junction FET utilizes an N-type channel and a P-type _____.

7. In an N-channel FET the majority carriers are _____.

8. In a P-channel FET the marjority carriers are _____.

9. The PN junction formed between the gate and source of a junction FET must always be _____-biased.

10. The input resistance of a junction FET is much higher than the input resistance of a bipolar transistor.

    A. True
    B. False

11. The current flowing through the channel of an FET is referred to as _____ current.

12. The current flowing through the channel of an FET can be effectively controlled by varying the reverse bias voltage applied to the gate and source lead.

    A. True
    B. False

13. The FET's gate-to-source voltage (V_{GS}) determines the size of the _____ region that is formed within the FET's channel.

14. A junction FET operates in the _____ mode.

15. A junction FET is biased so that an increase in V_{GS} will result in a _____ in I_D.

16. The value of $V_{GS(off)}$ for a given FET will be approximately equal to the FET's V_P value.

    A. True
    B. False

17. The transconductance of an FET is determined by dividing a small change in I_D by a small change in _____.

18. A P-channel FET is properly biased if its gate is positive with respect to its source.

    A. True
    B. False

19. When the arrow in a junction FET symbol points inward, the device utilizes a channel that is made of _____-type material.

20. A P-channel FET should be biased so that its drain is _____ with respect to its source.

The Insulated Gate FET

The gate and channel regions within a junction FET form a conventional PN junction and this junction is reverse-biased by connecting an external voltage between the FET's gate and source leads. This reverse bias voltage causes the FET to operate in the depletion mode and allows the device to have an extremely high input resistance. However, there is another type of FET which does not have a conventional PN junction that has to be reverse biased. This device uses a metal gate which is electrically insulated from its semiconductor channel by a thin oxide layer and is therefore referred to as an insulated gate FET (IGFET). However, this device is also known as a metal-oxide semiconductor FET (MOSFET).

Unlike the junction FET which operates in only the depletion mode, the IGFET (MOSFET) is designed to operate in one of two distinct modes. The IGFET may be either a depletion-mode device or an enhancement-mode device.

You will now see how insulated-gate FET's are constructed and you will learn their important electrical characteristics.

Depletion-Mode Devices

A cross-sectional view of a depletion-mode IGFET (MOSFET) is shown in Figure 7-5A. This device is formed by implanting an N-type channel within a P-type substrate. A thin insulating layer (silicon dioxide) is then deposited on top of the device; however the opposite ends of the N-type channel are left exposed so that wires or leads can be attached to the channel material. These two leads serve as the FET's source and drain. A thin metallic layer is then attached to the insulating layer so that it is directly over the N-type channel. This metal layer serves as the FET's gate and signals are applied to this metal gate through a suitable wire or lead. An additional lead is also connected to the substrate as shown.

It is important to note that the metal gate is insulated from the semiconductor channel by a layer of silicon dioxide. Therefore, the gate and the channel do not form a conventional PN junction. However, this insulated metal gate can be used to control the conductivity of the channel thus allowing the device to operate in a manner similar to that of a junction FET. In other words the insulated gate can deplete the N-type channel of majority carriers (electrons) when a suitable bias voltage is applied, even though a semiconductor junction does not exist between the gate and the channel.

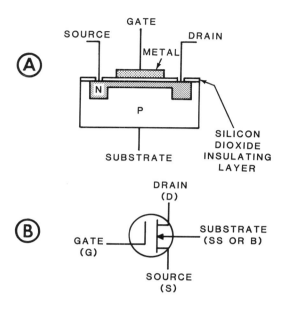

Figure 7-5 An N-channel depletion-mode IGFET and its schematic symbol.

The insulated gate FET shown in Figure 7-5A has an N-type channel; therefore its source and drain leads are biased in the same manner as the source and drain leads of an N-channel junction FET. In other words the drain is always made positive with respect to the source. The majority carriers (electrons) within the N-type channel therefore allows current to flow through the channel (from source to drain). This source-to-drain current (normally called drain current) is in turn controlled by a gate-to-source bias voltage just like a junction FET. When the gate-to-source voltage is equal to zero, a substantial drain current will flow through the device because a large number of majority carriers are present in the channel. If the gate is made negative with respect to the source, the channel becomes depleted of many of its majority carriers and drain current decreases just as it would in a junction FET. If this negative gate voltage is increased to a sufficiently high value, the drain current will drop to zero just like it does in the junction FET. There is one important difference between the operation of the N-channel insulated gate FET just described and the N-channel junction FET. The gate of the N-channel insulated gate FET can be made positive with respect to the FET's source. This must never occur in an N-channel junction FET because the PN junction formed by the gate and the channel would be forward biased.

The insulated gate FET can handle a positive gate voltage because the silicon dioxide insulating layer prevents any current from flowing through the gate lead. The FET's input resistance remains at a high value. Also, when a positive voltage is applied to the gate, more majority carriers (electrons) are drawn into the channel thus enhancing the conductivity of the channel. This is exactly opposite the action that takes place when the gate receives a negative voltage. A positive gate voltage can be used to increase the FET's drain current while a negative gate voltage will decrease the drain current.

Since a negative gate voltage is required to deplete the N-type channel insulated gate FET just described (and therefore reduce its drain current), this FET is called a depletion-mode device. Like the junction

FET described earlier (also a depletion-mode device) this FET conducts a substantial amount of drain current when its gate voltage is equal to zero. Therefore, all depletion-mode devices (either junction or insulated gate types) are said to be normally-conducting or normally-on when their gate voltages are zero.

The N-channel, depletion-mode IGFET just described is often represented by the schematic symbol shown in Figure 7-5B. Notice that the gate (G) lead is separated from the source (S) and drain (D) leads. Also the arrow on the substrate (often designated as SS or B) lead points inward to represent an N-channel device. Some FET's are constructed so that the substrate is internally connected to the source lead and a separate substrate lead is not used. In such a case, the symbol can be drawn so that it shows the substrate connected to the source.

A properly biased N-channel, depletion-mode IGFET is shown in Figure 7-6. Notice that this FET is biased in the same manner as an N-channel junction FET. The drain-to-source voltage (V_{DS}) must always be applied so that the drain is positive with respect to the source as shown. However, the gate-to-source voltage (V_{GS}) can be applied as shown or the polarity of V_{GS} can be reversed. In other words the gate can be made either negative or positive with respect to the source. Also notice that the substrate (B) has been externally connected to the source (S). The substrate is usually connected to the source (internally or externally) but this is not always the case. In some applications the substrate may be connected to the gate or to other points within the FET's respective circuit. These various circuit arrangements will be described later.

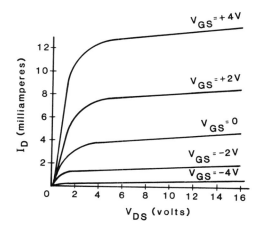

Figure 7-7 Typical drain characteristic curves for an N-channel depletion-mode IGFET.

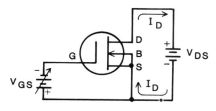

Figure 7-6 A properly biased N-channel depletion-mode IGFET.

The relationship between V_{GS}, V_{DS}, and I_D in a particular N-channel IGFET can be determined by examining the FET's drain characteristic curves. A typical set of curves are shown in Figure 7-7. Such curves could be plotted by using a circuit like the one shown in Figure 7-6. The curves are formed by adjusting V_{GS} to various values and then observing the relationship between I_D and V_{DS} for each V_{GS} value. If you compare these curves with the junction FET curves shown in Figure 7-3 you will find that they are similar. The basic difference is that positive as well as negative V_{GS} values are plotted in Figure 7-7 and only negative V_{GS} values are plotted in Figure 7-3.

The curves for the depletion-mode IGFET have the same general shape as the curves for the junction FET. In other words each curve rises rapidly and then levels off at a specific pinch-off voltage (V_p). The depletion-mode IGFET (like the junction FET) is usually operated above pinch-off where I_D is relatively constant with changes in V_{DS}. In this region the device may be used as a highly efficient voltage amplifier. However, there are some applications where the device is operated below pinch-off. Below pinch-off, the FET's drain current (I_D) varies over a wide range and at an almost linear rate as V_{DS} changes. This means that the resistance of the device can be varied over a wide range by an input gate voltage thus making the device useful as a voltage controlled resistor. Furthermore, the N-channel depletion mode IGFET is usually operated with a slightly negative gate-to-source bias voltage (V_{GS}) but in some cases V_{GS} may be equal to zero. This means that an AC input signal voltage can be applied to the FET's gate (in series with V_{GS}) so that the resulting V_{GS} is effectively made to vary in value but remain negative or effectively vary between positive and negative values.

Depletion-mode IGFET's may also be constructed in a manner which is exactly opposite to the N-channel device shown in Figure 7-5. In other words they may utilize P-type channels which are implanted within N-type substrates. Such devices are generally referred to as P-channel depletion-mode IGFET's. These P-channel devices operate in basically the same manner as the N-channel devices just described, however, the majority carriers within the P-type channel are holes instead of electrons. Furthermore, the drain lead of the P-channel device must be made negative with respect to its source lead so that its drain-to-source voltage (V_{DS}) is exactly opposite to the V_{DS} applied to the N-channel device in Figure 7-6. This means that the drain current (I_D) must also flow in the opposite direction. However, the gate may be negative or positive with respect to the source as with the N-channel device. The drain characteristic curves for the P-channel depletion-mode IGFET are therefore similar to the curves for an N-channel device but of course the polarities of the voltages and currents involved are reversed.

The schematic symbol for a P-channel, depletion-mode IGFET is shown in Figure 7-8. Notice that the only difference be- tween this symbol and the symbol for an N-channel device is the direction of the arrow. In this symbol the arrow points outward to indicate that a P-type channel is used.

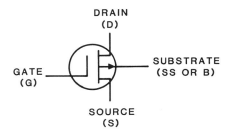

Figure 7-8 Schematic symbol for a P-channel depletion-mode IGFET.

Both N-channel and P-channel, depletion-mode IGFET's may be constructed in a symmetrical manner (like the junction FET) so that their source and drain leads may be interchanged. However, these devices are sometimes constructed so that their gates are offset from their drain regions. This reduces the capacitance between the gate and drain and improves the operation of the device in certain applications. When using this type of FET, the source and drain leads should not be interchanged.

One widely used variation of the depletion mode device is a dual gate unit. A dual gate IGFET (MOSFET) has two gate elements each of which can be used separately to control the drain current. Such devices are used as mixers or as gain controlled amplifiers in radio and TV receivers.

Figure 7-9 shows the symbol for an N-channel dual gate IGFET (MOSFET). Note the back-to-back zener diodes connected between each gate and the source. These are used to protect against electrostatic charges that can damage the device. The substrate is internally connected to the source. Each gate has equal control over the drain current.

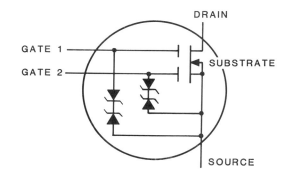

Figure 7-9 A dual gate depletion mode IGFET with gate protection diodes.

Enhancement-Mode Devices

As explained previously, the depletion-mode IGFET is a normally-on device and therefore conducts a substantial drain current (I_D) when its gate-to-source voltage (V_{GS}) is zero. Although this type of FET is useful in many applications, there are also certain applications where a normally-off device is required. In other words it is often useful to have a device that conducts negligible I_D when V_{GS} is equal to zero but will allow I_D to flow when a suitable V_{GS} value is applied.

An IGFET which will function as a normally-off device can be constructed as shown in Figure 7-10A. This device is similar to the depletion-mode IGFET but it does not have a conducting channel which is imbedded in the substrate material. Instead, the device has source and drain regions which are diffused separately into the substrate. The device in Figure 7-10A has a P-type substrate and N-type source and drain regions; however, the exact opposite arrangement can also be used. The metal gate is separated from the substrate material by a silicon dioxide insulating layer as shown and the various leads are attached to the device so that it has basically the same lead arrangement as the depletion-mode device.

The device shown in Figure 7-10A must be biased so that its drain is made positive with respect to its source. With only a drain-to-source voltage (V_{DS}) applied, the FET does not conduct a drain current (I_D). This is because no conducting channel exists between the source and drain regions. However, this situation can be changed with the application of a suitable gate voltage. When the gate is made positive with respect to the source; electrons are drawn toward the gate. These electrons accumulate under the gate where they create an N-type channel which allows current to flow from source to drain. When this positive gate voltage increases, the size of the channel also increases, thus, allowing even more drain current to flow.

The action just described is similar to the action that takes place in a charging capacitor. The metallic gate and the substrate act like the upper and lower plates of a capacitor and the insulating layer acts like a dielectric. The positive gate voltage simply causes the capacitor to charge and a negative charge builds up on the substrate side of the capacitor. The positive gate voltage effectively induces an N-type channel between the source and drain regions which sustains a drain current. Furthermore, an increase in gate voltage tends to enhance the drain current. For these reasons the device in Figure 7-10A is commonly referred to as an N-channel, enhancement-mode IGFET (MOSFET).

The gate of the N-channel, enhancement mode IGFET can also be made negative with respect to its source. However, a negative gate voltage will not affect the operation of the FET since it is a normally-off (non-conducting) device. In other words, the FET's drain current is normally equal to zero and therefore cannot be further reduced by the application of a negative gate voltage.

The schematic symbol for an N-channel, enhancement-mode IGFET is shown in Figure 7-10B. Notice that this symbol is similar to the symbol for the N-channel depletion-mode IGFET shown in Figure 7-5B. The only difference is the use of an interrupted line instead of a solid line to interconnect the source, drain, and substrate regions. The solid line is used to identify the normally-on condition of the depletion-mode device while the interrupted line is used to identify the normally-off condition of the enhancement-mode device. In each symbol the arrow points inward to indicate that each device has an N-type channel.

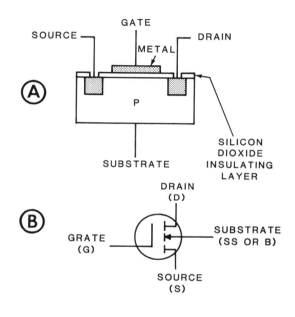

Figure 7-10 An N-channel enhancement-mode IFGET and its schematic symbol.

A properly biased N-channel, enhancement-mode IGFET is shown in Figure 7-11. Notice that the FET's drain is made positive with respect to its source by the drain-to-source voltage (V_{DS}). However, the gate is also made positive with respect to the source by the gate-to-source voltage (V_{GS}). Only when V_{GS} increases from zero and applies a positive voltage to the gate, will a substantial amount of drain current (I_D) flow. The substrate is normally connected to the source as shown, but in special applications, the substrate and source may be at different potentials.

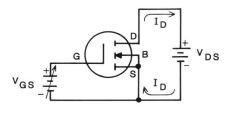

Figure 7-11 A properly biased N-channel enhancement-mode IGFET.

The relationship between V_{GS}, V_{DS}, and I_D in a particular N-channel, enhancement-mode IGFET can be determined by examining the FET's drain characteristic curves. A typical set of these curves is shown in Figure 7-12. Such curves are usually plotted by biasing the FET as shown in Figure 7-11. Then V_{GS} is adjusted to various values and the relationship between I_D and V_{DS} is observed for each V_{GS} value. These curves are similar to the curves for an N-channel, depletion-mode IGFET but only positive values of V_{GS} are plotted instead of both positive and negative values. Notice that for each higher positive value of V_{GS}, I_D rises to a correspondingly higher value and then levels off.

Although it is not apparent in Figure 7-12, V_{GS} must exceed a certain threshold voltage (usually one volt or more) before the N-type channel induced within the FET is great enough to support a usable current. Any V_{GS} value below this threshold cannot cause the

FET to conduct and the device will effectively act as if its gate voltage was equal to zero. The N-channel enhancement-mode IGFET therefore normally operates with a positive gate bias that is greater than its threshold voltage. With proper biasing the enhancement-mode device makes an excellent switch. The device can be turned on by a sufficiently high gate voltage and turned off when the gate voltage drops below the threshold level. The inherent threshold of the device therefore provides a highly desirable region of noise immunity which prevents low or intermediate input voltages (below threshold) from falsely triggering the device on. This characteristic makes the enhancement mode device ideal for digital applications involving logic and switching functions. The enhancement mode IGFET is the basic component used in many large scale digital integrated circuits.

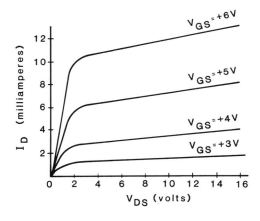

Figure 7-12 Typical drain characteristic curves for an N-channel enhancement-mode IGFET.

An enhancement-mode IGFET may also be constructed in a manner that is exactly opposite the device shown in Figure 7-10A. In other words the device may have P-type source and drain regions which are implanted into an N-type substrate. This type of FET must be operated with a negative gate voltage so that holes (instead of electrons) are attracted toward the gate to form a P-type channel. Such a device is referred to as a P-channel, enhancement-mode IGFET (MOSFET). The P-channel device functions in basically the same manner as the N-channel device even though holes instead of electrons are used to support drain current through the device. The P-channel device requires bias voltages (V_{GS} and V_{DS}) that are opposite to those shown in Figure 7-11. Also, the drain characteristic curves for the P-channel device have the same general shape as the curves for the N-channel device although the polarities of the voltages and currents involved are reversed.

The symbol for a P-channel, enhancement-mode IGFET is shown in Figure 7-13. This symbol closely resembles the symbol for the N-channel device. The only difference is the direction of the arrow which in this case points outward to identify the P-type channel that is induced within the device. The enhancement-mode device is usually constructed in a symmetrical manner just like the junction FET. This means that the source and drain leads can usually be reversed or interchanged.

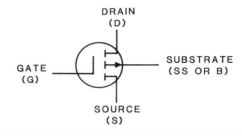

Figure 7-13 Schematic symbol for a P-channel enhancement-mode IGFET.

CMOS

CMOS devices (Complementary-Metal-Oxide-Semiconductor) or components were developed using a new technology. This new technology is a logical expansion of the MOS technology that followed the development of TTL logic (bipolar components). As you know, bipolar devices (PNP and NPN transistors) were combined to make a circuit called a complementary pair. MOS devices were also developed as P-type and N-type components and then externally combined to form a complementary pair or a complementary amplifier. CMOS technology combines the N-type and P-type FETs together into a single unit that is a complementary pair. This is shown in Figure 7-14.

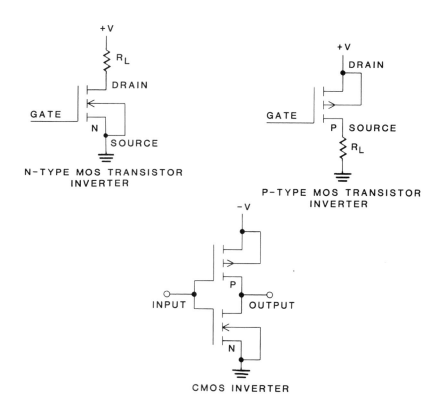

Figure 7-14 Schematic diagrams of inverter circuits.

The combining of N-type and P-type FETs within a single unit requires less space and reduces power supply requirements. This in turn reduces the power dissipated and the noise generated internally.

Any of the functions that can be performed by conventional FETs can be accomplished with CMOS devices. However, CMOS components operate at slower speeds than the P-type or N-type FETs. This is because the propogation delay (time it takes a signal to travel through a device) is longer in CMOS devices. Remember that the FET operates at slower speeds than a bipolar device. The FET is also limited to lower power applications than the bipolar components. CMOS devices are slightly slower than FETs.

In addition to slow operating speeds (approximately 5 million changes per second), and low power applications, the CMOS device is more susceptible to damage from static electricity than either a TTL or FET component. Static electricity may build up on your fingers to a high enough potential to destroy a CMOS component. Therefore, special handling and packaging are required. Most CMOS devices are shipped with the pins inserted into conductive foam or wrapped with a shorting band or clip. When CMOS devices are used, it is a good idea to keep the pins shorted until the circuit is completely constructed. It is also helpful to wear a static discharge band while handling CMOS devices. As with other components, it is not a good idea to remove or insert CMOS components into an activated circuit.

This section of the text is designed only as an introduction to CMOS technology. A detailed discussion on CMOS components and circuits can be found in Heathkit/Zenith's Digital Techniques course. Heathkit/Zenith's CMOS Digital Techniques course is written from an applications point of view. If you are not already familiar with the terms and theories used in digital electronics the CMOS course should follow the Digital Techniques course in your study plan.

Safety Precautions

When using any IGFET certain precautions must be observed. As with any solid-state component it is necessary to check the manufacturer's maximum ratings so that the device is not damaged by excessively high operating voltages or currents. However, it is particularly important to observe the FET's maximum allowable gate-to-source voltage (V_{GS}). An IGFET (MOSFET) can accept only a limited range of V_{GS} values because of the extremely thin silicon dioxide insulating layer that separates its gate and channel. If V_{GS} is increased too much, the thin insulating layer will be punctured and the device will be ruined. In fact, the insulating layer is so sensitive that it can even be damaged by static charges that build up on the FET's leads. For example, the electrostatic charges on your fingers can be transferred to the FET's leads while handling the device or when mounting it in a circuit. The device could therefore be ruined before it is used.

To avoid this type of damage, manufacturers usually ship these devices with their leads shorted together so that static charges cannot build up between the leads. The leads may be wrapped with a shorting wire, or inserted within a shorting ring, or pressed into a conducting foam material, or simply taped together.

These shorting devices should not be removed until the FET is completely installed in its respective circuit. Components should not be installed or removed when power is applied to the circuit.

Most modern IGFETs are protected by zener diodes which are electrically connected between each insulated gate and the transistor's source inside the device. These diodes offer protection against static discharge and in-circuit transients without the need for external shorting mechanisms.

Devices which do not include gate-protection diodes can be handled safely if the following basic precautions are taken:

SAFETY NOTES

1. Prior to connection in a circuit, all leads of the device should be kept shorted together either by the use of a metal shorting ring attached to the device by the manufacturers, or by their insertion into some kind of conductive material. Aluminum foil is sometimes used, since it can be readily torn away after the device is installed.

2. When devices are removed by hand from their carriers, the hand being used should be grounded by any suitable means, for example, with a metallic wristband.

3. Tips of soldering irons should be grounded before soldering the device.

4. The device should never be inserted into, or removed from circuits with power on.

These precautions apply to both depletion and enhancement-mode IGFETS (MOSFETs) as well as other sensitive components.

21. An insulated gate FET has a metallic gate which is electrically insulated from its semiconductor _____.

22. An IGFET may be designed to operate in the depletion mode or the enhancement mode.

    A. True
    B. False

23. A depletion-mode IGFET can be either a _____-channel or _____ -channel device.

24. A depletion-mode IGFET operates similar to a junction FET but it does not have a PN _____.

25. The majority carriers within an N-channel, depletion-mode IGFET are _____.

26. The N-channel, depletion-mode IGFET is always biased so that its drain is _____ with respect to its source.

27. When the N-channel, depletion-mode IGFET's gate-to-source voltage is equal to zero, drain current will flow through the device.

    A. True
    B. False

28. When the gate of an N-channel, depletion-mode IGFET is made increasingly negative with respect to its source, the FET's drain current will _____.

29. If the gate of an N-channel, depletion-mode IGFET is made increasingly positive with respect to its source, the FET's drain current will _____.

30. The relationship between V_{GS}, V_{DS}, and I_D in a depletion-mode IGFET can be determined by examining the FET's _____ _____ curves.

31. A P-channel, depletion-mode IGFET requires bias voltages that are exactly _____ to those used with an N-channel device.

32. The majority carriers within a P-channel, depletion-mode IGFET are _____.

33. Depletion-mode IGFET's are often referred to as normally-_____ devices.

34. An enhancement-mode IGFET conducts only a negligible drain current when V_{GS} is equal to _____.

35. The enhancement-mode IGFET is said to be a normally-_____ device.

36. In an N-channel, enhancement-mode IGFET, a positive gate voltage will attract _____ toward the gate to form an N-type channel.

37. In order to conduct drain current an N-channel enhancement-mode IGFET must have a _____ gate voltage.

38. The metallic gate and the substrate in an enhancement-mode IGFET act like a _____.

39. The drain current in an N-channel, enhancement-mode IGFET can be reduced below its normal value by application of a negative gate voltage.

    A. True
    B. False

40. The N-channel, enhancement-mode IGFET will normally be biased so that its drain is _____ with respect to its source.

41. The drain characteristic curves for the N-channel, enhancement-mode IGFET resemble the curves for an N-channel, depletion mode IGFET, however the enhancement-mode curves are only plotted for _____ values of V_{GS}.

42. In order for an enhancement-mode IGFET to conduct a substantial drain current, V_{GS} must exceed a certain _____ voltage.

43. In a P-channel, enhancement-mode IGFET, a negative gate voltage will attract _____ toward the gate to form a P-type channel.

44. When using any type of insulated gate FET it is necessary to insure that static charges do not build up on the component leads so that the highly sensitive _____ layer will not be damaged.

45. Some IGFETs are protected from electrostatic charges by internally connected _____ _____.

FET Circuit Arrangements

Like bipolar transistors, FET's are used primarily to obtain amplification and like bipolar transistors, FET's can be connected in three different circuit arrangements. These three configurations are commonly referred to as common-source, common-gate, and common-drain circuits. These configurations are valid for both JFETs and IGFETs.

Common-Source Circuits

The common-source circuit is the most widely used FET circuit arrangement. This circuit configuration is comparable to the common-emitter (bipolar-transistor) circuit arrangement that was described in a previous unit. A basic common-source configuration is shown in Figure 7-15. Notice that the input signal is applied between the gate and source leads of the FET and the output signal appears between the drain and source leads. The source is therefore common to both input and output.

An N-channel junction FET is shown in Figure 7-15 and it is therefore biased so that its gate is negative with respect to its source. The gate-to-source bias voltage is provided by an external voltage source (designated as V_{GG}) which is in series with a resistor (R_G). Therefore, V_{GG} is not applied directly to the gate and source leads. However, the FET's gate-to-channel PN junction is reverse-biased so that only an insignificant leakage current can flow through the gate lead and R_G. The voltage across R_G is almost zero and

the full value of V_{GG} is effectively placed across the gate and source leads. Normally R_G has a high value of resistance (often more than 1 megohm) so that the resistance seen at the input of the circuit will remain high. A low value of resistance could reduce the input resistance of the circuit since R_G is effectively in parallel with the FET's high input (gate-to-source) resistance as far as the input signal is concerned.

The external bias voltage (V_{GG}) is adjusted so that a specific value of drain current (I_D) will flow through the FET and the device will operate within its pinch-off region. The value of I_D is also controlled (but to a lesser extent) by the external voltage source V_{DD}. This voltage source supplies the necessary drain-to-source operating voltage for the FET but another resistor (R_L) is inserted between V_{DD} and the FET's drain lead. The drain current flowing through the FET therefore flows through R_L, thus, causing a voltage drop across R_L. A portion of the voltage (V_{DD}) is also dropped across the FET. The FET and R_L form a series voltage divider.

Since the FET is controlled by an input voltage (not an input current) the common-source arrangement is used to obtain voltage amplification. For example, an AC input voltage will alternately aid and oppose the input bias voltage (V_{GG}) so that the FET's gate-to-source voltage will vary with the changes in input voltage. This will cause the FET to alternately conduct more and less drain current. The FET becomes a variable resistor in series with the fixed resistor (R_L)

and the voltage. As the FET conducts more and less drain current, its drain-to-source voltage varies accordingly to produce an output voltage that changes in response to the input voltage. However; by making R_L relatively large (often more than 10 k ohms) and by biasing the FET so that its drain-to-source resistance is also high, the changes in drain current (even when small) will produce an output signal voltage that is much higher than the input signal voltage.

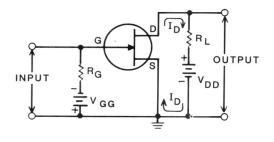

Figure 7-15 Basic common-source circuit.

In addition to providing voltage amplification, the common-source circuit has another desirable feature. Due to the extremely high gate-to-source resistance of the FET, the circuit has a very high input resistance even though input resistor R_G is used. This means that the common-source circuit will usually have a minimum loading effect on its input signal source. Common-source circuits are therefore widely used in digital or computer applications where a number of circuit inputs must often be connected to the output of a single circuit without affecting its operation.

The output resistance of the common-source circuit is lower than its input resistance because of the lower source-to-drain resistance of the FET. However, the circuit still has a moderately high output resistance, since the FET is usually biased to conduct a relatively low drain current which is often no more than a few milliamperes.

The common-source circuit may also be used to amplify both low frequency and high frequency AC signal voltages as well as a wide range of DC signal voltages. For example, this circuit is often used in the input RF amplifier stage of radio receivers to amplify the wide range of high frequency input signals which can have large amplitude variations. The circuit is also used in electronic test instruments such as solid-state voltmeters or multimeters. In these applications it must amplify a wide range of DC and AC voltages and at the same time present a high input resistance to prevent undesirable loading of the circuits being tested.

Although the common-source circuit shown in Figure 7-15 is formed with an N-channel junction FET, the same basic circuit can be formed with a P-channel JFET if the polarities of the bias voltages are reversed. Also, this basic circuit can be formed with depletion-mode or enhancement-mode IGFET's. When these insulated-gate devices are used, the additional substrate leads are generally connected to their respective source leads or to circuit ground. Also, it is important to note that depletion-mode

IGFET's can accept positive and negative gate voltages and may be operated with zero gate bias, while enhancement-mode devices require a gate bias voltage in order to conduct. This means that slightly different biasing arrangements must be used to accommodate the various FET types when the basic common-source arrangement is used.

Common-Gate Circuits

The common-gate circuit may be compared to the common-base bipolar transistor circuit because the electrode which has primary control over the FET's conduction (its gate) is common to the circuit's input and output. A basic common-gate circuit is shown in Figure 7-16. An N-channel JFET is used in this circuit and it therefore requires the same basic operating voltages as the N-channel device in the common-source circuit previously described. However, in this circuit the input gate-to-source bias voltage is provided by voltage V_{SS} and resistor R_S while the output portion of the circuit is biased by the voltage V_{DD} and resistor R_L.

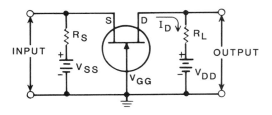

Figure 7-16 Basic common-gate circuit.

An AC input voltage will effectively vary the FET's gate-to-source voltage and cause variations in its conduction. The FET's drain current flows through R_L and variations in this current produce voltage changes across the FET which follow the input signal. The voltage developed across the drain and gate leads of the FET serves as the output.

Like the common-source circuit, the common-gate circuit provides voltage amplification; however the voltage gain of the common-gate arrangement is lower. The input resistance of the circuit is also low since current flows through the input source lead; however the output resistance of the circuit is relatively high. This makes the common-gate circuit suitable in applications where a low resistance source must supply power to a high resistance load. When inserted between the source and load, the common-gate circuit effectively matches the low and high resistance to ensure an efficient transfer of power.

Although it has a low voltage gain, the common-gate circuit is often used to amplify high frequency AC signals. This is because of the circuit's low input resistance and because the circuit inherently prevents any portion of its output signal from feeding back and interfering with its input signal. The circuit is therefore inherently stable at high frequencies and additional components are not required to prevent interference between input and output signals. Such stabilizing components are often required with the common-source circuit arrangement to insure reliable operation.

The common-gate circuit may also be formed with a P-channel JFET or with depletion-mode or enhancement-mode IGFET's. However, when an IGFET is used, the substrate lead is usually connected directly to the gate or to circuit ground and additional components may be required to properly bias the device.

Common-Drain Circuits

The common-drain circuit is similar to the common-collector bipolar transistor circuit. This basic FET circuit arrangement is shown in Figure 7-17. The N-channel JFET in this circuit receives its gate-to-source bias voltage from V_{GG} and R_G. The output (drain-to-source) portion of the FET is biased by V_{DD} and R_L. The input signal voltage is effectively applied betwen the FET's gate and drain even though it appears to be applied to the gate and source leads. The output signal is effectively developed across the FET's source and drain even though it appears to be taken directly from R_L. This is because voltage source V_{DD} is effectively a short as far as the input and output signals are concerned. Therefore, V_{DD} effectively grounds the FET's drain and makes it common to both input and output signals.

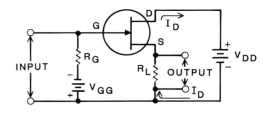

Figure 7-17 Basic common-drain or source follower circuit.

The common-drain circuit cannot provide voltage amplification. The voltage appearing at the source of the FET tends to track or follow the voltage at the gate and for this reason the circuit is often referred to as a source-follower. However the output source voltage is always slightly less than the input gate voltage, thus, causing the circuit to have a voltage gain that is slightly less than one.

The input resistance of the common-drain circuit is extremely high. In fact, this circuit has a higher input resistance than the common-source or common-gate configurations. However, the output resistance of this circuit is low. This makes the common-drain circuit suitable for coupling a high resistance source to a low resistance load so that an efficient transfer of power can take place.

The common-drain configuration can also be formed with P-channel JFET's as well as depletion-mode and enhancement-mode IGFET's.

46. The most widely used FET circuit arrangement is the _____-_____ circuit.

47. Only N-channel JFET's can be used to form the three basic FET circuit arrangements.

    A. True
    B. False

48. The FET circuit arrangement that has the highest voltage gain is the _____-_____ circuit.

49. The FET circuit arrangement that inherently provides the most stable operation at high signal frequencies is the _____-_____ circuit.

50. The common-drain arrangement provides a voltage gain of less than _____.

51. The common-drain arrangement is sometimes referred to as a _____-_____.

52. To match a high impedance source to a low impedance load the common _____ configuration should be used.

53. To match a low impedance source to a high impedance load the common _____ configuration should be used.

54. When compared to bipolar transistors, FETs have a _____ input impedance.

Unit Summary

The field effect transistor can provide amplification of electronic signals like a conventional bipolar transistor. In fact, the FET is even more efficient than a bipolar transistor in certain applications. However, the FET is considered a voltage controlled device while a bipolar transistor is considered a current operated device.

The operation of the bipolar transistor depends on the movement of both majority and minority carriers, but the FET operates only with majority carriers. The majority carriers flow through a semiconductor channel which is implanted within an oppositely doped substrate. The opposite ends of the channel are referred to as the FET's source and drain and may be compared to the emitter and collector in a bipolar transistor. The substrate itself is used to control the movement of majority carriers through the channel and it is referred to as the FET's gate. This gate may be compared to the base of a bipolar transistor. The device just described is more appropriately referred to as a junction FET or JFET since the substrate (gate) and the semiconductor channel form a PN junction.

Junction FET's may be constructed with an N-type channel and a P-type substrate or with a P-type channel and an N-type substrate. These devices are generally referred to as just N-channel or P-channel junction FET's respectively. The substrate (gate) to channel junction of the FET must always be reverse-biased so that a depletion region will form within the channel. By adjusting this reverse voltage, the size of the depletion region can be controlled. Since the size of the depletion region in turn controls the movement of the majority carriers through the channel, the gate-to-channel voltage can be used to effectively control the conduction of the FET.

The gate-to-channel junction is usually reverse-biased by applying the bias voltage between the FET's gate and source leads. In addition to this gate-to-source bias voltage, the FET also requires a source-to-drain bias voltage to force the majority carriers to move through the channel. The majority carriers will be electrons in an N-channel device and holes in a P-channel device, but in either case only electrons leave the channel to flow through the external bias voltage source and return to the channel to complete the circuit. This external flow of electrons is referred to as the FET's drain current.

The FET's gate-to-source voltage controls the depletion region within its channel and the size of this region in turn controls the FET's drain current. The junction FET is therefore said to operate in the depletion-mode.

The junction FET represents one basic type of FET. A second type, known as an insulated gate FET (IGFET) is also used. The insulated gate FET, also called a MOSFET, is constructed in a manner similar to the junction device but it utilizes a metallic gate which is electrically insulated from its semiconductor channel. The substrate is used only to support the entire structure.

The substrate may be internally connected to the source or a separate electrical connection to the substrate may be provided. Insulated gate FET's are also available as N-channel and P-channel devices. Furthermore, they can be designed to operate in either the depletion mode like the junction FET or they can operate in the enhancement mode. The depletion-mode device operates similar to the junction FET, but its gate can be made either negative or positive with respect to its source. This is true if the device has either an N-type or a P-type channel. Furthermore, the depletion-mode device conducts drain current with a zero gate-to-source voltage just like the junction FET and is a normally-on device.

The enhancement-mode device does not conduct drain current when its gate-to-source voltage is zero. Therefore, this type of insulated gate FET is often referred to as a normally-off device. A gate-to-source voltage must be applied to enhance conduction within the device. The gate of the enhancement-mode FET can be made either positive or negative with respect to its source. However, only one polarity of input gate voltage can be used to turn on the device and control its conduction. In general, N-channel devices require positive gate voltages and P-channel devices require negative gate voltages, to turn on.

Either type of FET can be connected in a common-source configuration so that the input signal is applied between its gate and source, and the output signal voltage is developed between its drain and source. In this configuration the FET's reversed biased gate-to-channel junction gives the FET and the circuit a high input resistance. This arrangement also provides a substantial voltage gain. However the FET can be connected in a common-gate arrangement which has a somewhat lower voltage gain but provides more stable operation at high frequencies. A third basic configuration is also possible which is known as a common-drain or source-follower circuit. This arrangement provides no voltage gain but offers a very high input resistance and a low output resistance and can be used to insure maximum transfer of power from a high resistance source to a low resistance load.

Unit 8
Thyristors

Contents

Introduction

The term thyristor defines a broad range of solid-state components which are used as electronically controlled switches. Each of these devices can switch between a conducting (on) state and a nonconducting (off) state to effectively pass or block electrical current. Furthermore, some thyristors are capable of switching currents flowing in one direction while others can switch currents flowing in either direction.

Thyristors are widely used in applications where DC and AC power must be controlled. These devices are often used to apply a specific amount of power to a load or to completely remove it from the load. However, they are also used to regulate or adjust the amount of power applied to a specific load. For example, a thyristor could be used to simply turn an electric motor on or off or it could be used to adjust the speed or torque of the motor over a wide operating range.

Thyristors should not be confused with bipolar transistors or field effect transistors (FET's). Although it is true that transistors and FET's can be used as electronic switches, these devices are not as efficient and they do not have the power handling capability of thyristors. Thyristors are devices that are used expressly for the purpose of controlling electrical power while transistors and FET's are primarily used to provide amplification.

A variety of thyristors are now available but many of these devices have similar or related characteristics. Most applications which involve power control are handled with a few basic components. The thyristors that are most widely used are the silicon controlled rectifier (SCR), the bidirectional triode (TRIAC), the bidirectional trigger diode (DIAC), the unijunction transistor (UJT), and the programmed unijunction transistor (PUT). You will examine each of these devices as you proceed through this unit. You will also demonstrate the operation of the silicon controlled rectifier and the unijunction transistor.

When you have completed this unit on thyristors you should be able to:

1. Describe the conditions necessary to turn on, or turn off, a silicon controlled rectifier.

2. Explain the difference between a silicon controlled rectifier's forward breakover and reverse breakdown voltage.

3. Name two applications of the silicon controlled rectifier.

4. Describe the conditions necessary to turn on, or turn off, a bidirectional triode thyristor.

5. Name two applications of the bidirectional triode thyristor.

6. Describe the basic operation of the bidirectional trigger diode.

7. Name the most important application of the bidirectional trigger diode.

8. Describe the conditions required to turn on a unijunction transistor.

9. Explain how a unijunction transistor exhibits a negative resistance once it is turned on.

10. Name two applications of the unijunction transistor.

11. Describe the difference between an ordinary unijunction transistor and a programmed unijunction transistor.

12. From a schematic point out the:

    SCR

    TRIAC

    DIAC

    UJT

    PUT

Silicon Controlled Rectifiers

The silicon controlled rectifier is the most popular member of the thyristor family. This device is generally referred to as an SCR. Unlike the bipolar transistor which has two junctions and provides amplification, the SCR has three junctions and is used as a switch. As its name implies, the device is basically a rectifier which conducts current in only one direction. However, the device can be made to conduct (turn on) or stop conducting (turn off) and therefore provide a switching action that can be used to control electrical current.

Let's now take a close look at the SCR. First we will examine its basic construction and operation, and then we will consider its important electrical characteristics and basic applications.

Basic Construction and Operation

An SCR is a solid-state device which has four alternately doped semiconductor layers. The device is almost always made from silicon, but germanium has been used. The SCR's four layers are often formed by a diffusion process but a combined diffusion-alloyed method is also used.

A simplified diagram of an SCR is shown in Figure 8-1A. As shown, the SCR's four (PNPN) layers are sandwiched together to form three junctions. However, leads are attached to only three of the four layers. These three leads are referred to as the anode, cathode, and gate.

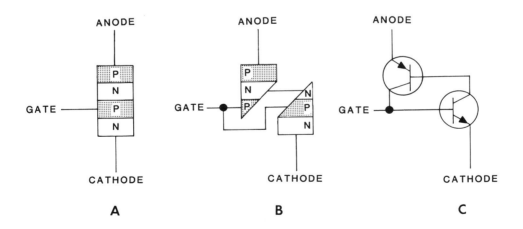

Figure 8-1 The SCR and its equivalent circuits.

The manner in which an SCR operates cannot be easily determined by examining the four-layer structure shown in Figure 8-1A. However, the SCR's four layers can be divided so that two three-layer devices are formed as shown in Figure 8-1B. The two devices obtained are effectively PNP and NPN transistors and when interconnected, as shown, they are equivalent to the four layer device shown in Figure 8-1A. These two transistors can also be represented by their schematic symbols as shown in Figure 8-1C.

We will now bias the equivalent circuit in Figure 8-1C just like we would bias the actual SCR shown in Figure 8-1A. First, we will make the anode of the circuit positive with respect to the cathode but we will leave the gate open. Under these conditions the NPN transistor will not conduct because its emitter junction will not be subjected to a forward bias voltage which can produce a base current. This will, in turn, cause the PNP transistor to turn off because the NPN transistor will not be conducting and therefore will not allow a base current to flow through the emitter junction of the PNP transistor. The equivalent SCR circuit will not allow current to flow from its cathode to its anode under these conditions.

If the gate of the equivalent SCR circuit is momentarily made positive with respect to the cathode, the emitter junction of the NPN transistor will become forward biased and this transistor will conduct. This will in

turn cause a base current to flow through the PNP transistor which will cause this transistor to conduct. However, the collector current flowing through the PNP transistor now causes base current to flow through the NPN transistor. The two transistors therefore hold each other in the on or conducting state, thus, allowing current to flow continuously from the cathode to the anode of the circuit. It is important to note that this action takes place even though the gate voltage is applied only for a moment. This momentary gate voltage causes the circuit to switch to the on or conducting state and the circuit will remain in that state even though the gate voltage is removed.

In order to switch the equivalent SCR circuit back to its off or nonconducting state, it is necessary to reduce its anode-to-cathode voltage to almost zero. This will cause both transistors to turn off and remain off until the gate voltage is again applied.

The SCR in Figure 8-1A operates just like the equivalent circuit in Figure 8-1C. In other words the SCR can be turned on by a positive input gate voltage and must be turned off by reducing its anode-to-cathode voltage. When the SCR is turned on and is conducting a high cathode-to-anode current, the device is said to be conducting in the forward direction. If the polarity of the cathode-to-anode bias voltage was reversed, the device would conduct only a small leakage current which would flow in the reverse direction.

The SCR is usually represented by the schematic symbol shown in Figure 8-2. Notice that this symbol is actually an ordinary diode symbol with an added gate lead. The circle surrounding the diode may or may not be used and the SCR's anode, gate and cathode leads may or may not be identified. When the leads are identified, they are usually represented by the letters A, G and K as shown.

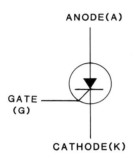

Figure 8-2 Schematic symbol of an SCR.

A properly biased SCR is shown schematically in Figure 8-3. Notice that a switch (S) is used to apply or remove the input gate voltage which is obtained from a voltage source and resistor R_G. This resistor is used to limit the gate current (I_G) to a specific value. The SCR's anode-to-cathode voltage is provided by another voltage source but a series load resistor (R_L) is also used to limit the SCR's cathode-to-anode current to a safe value when the device is turned on. Without this resistor, the SCR would conduct a very high cathode-to-anode current

(also referred to as anode current or I_A) and could be permanently damaged.

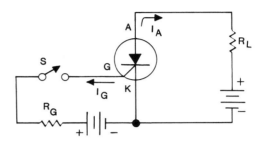

Figure 8-3 A properly biased SCR.

V-I Characteristics

A better understanding of SCR operation can be obtained by examining the voltage-current (V-I) curve shown in Figure 8-4. This curve shows the V-I characteristics of a typical SCR. Such a curve is plotted by varying the SCR's cathode-to-anode voltage over a wide range while observing the SCR's anode current. The corresponding values are then plotted and a continuous curve is formed. The SCR is first biased in the forward direction while its gate is open as shown in Figure 8-4. The SCR's cathode-to-anode voltage is designated as V_F at this time. The curve shows that as V_F increases from zero, the SCR conducts only a small forward current (designated as I_F) which is due to leakage. As V_F continues to increase,

I_F remains very low and almost constant but eventually a point is reached where I_F increases rapidly and V_F drops to a low value (note the horizontal dotted line). The V_F value required to trigger this sudden change is referred to as the forward breakover voltage. When this value of V_F is reached the SCR simply breaks down, and conducts a high I_F which is limited only by the external resistance in series with the device. The SCR switches from the off state to the on state at this time. The drop in V_F occurs because the SCR's resistance drops to an extremely low value and most of the source voltage appears across the series resistor.

When the SCR is in the on state, only a slight increase in V_F is required to produce a tremendous increase in I_F (the curve is almost vertical and straight). Furthermore, the SCR will remain in the on state as long as I_F remains at a substantial value. Only when I_F drops below a certain minimum value, will the SCR switch back to its off state. This minimum value of I_F which will hold the SCR in the on state is referred to as the SCR's holding current and is usually designated as I_H. As shown in Figure 8-4, the I_H value is located at the point where breakover occurs (just to the left of the horizontal dotted line).

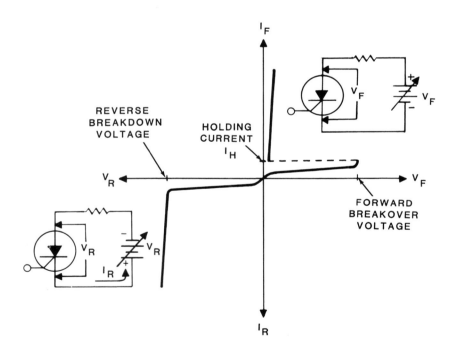

Figure 8-4 V-I characteristics of a typical SCR with the gate open.

When a reverse voltage is applied to the SCR as shown in Figure 8-4, the device functions in basically the same manner as a reverse-biased PN junction diode. As the reverse voltage (V_R) across the SCR increases from zero, only a small reverse current (I_R) will flow through the device due to leakage. This current will remain small until V_R becomes large enough to cause the SCR to breakdown. Then I_R will increase rapidly if V_R increases even slightly above the breakdown point (the curve is almost vertical and straight). The reverse voltage (V_R) required to breakdown the SCR is referred to as the SCR's reverse breakdown voltage. If too much reverse current is allowed to flow through the SCR after breakdown occurs, the device could be permanently damaged. However, this situation is normally avoided because the SCR is usually subjected to operating voltages which are well below its breakdown rating.

The V-I curve in Figure 8-4 shows the relationship between V_F and I_F when the SCR's gate is open. In other words no voltage is applied to the SCR's gate and no gate current is flowing through the device. The curve in Figure 8-4 could therefore be labeled to indicate that gate current is equal to zero.

When the gate is made positive with respect to the cathode, gate current will flow and the SCR's forward characteristics will be affected. The changes that will take place in the SCR's forward characteristics are graphically represented in Figure 8-5. In this figure, three V-I curves are plotted to show how changes in gate current (designated as I_G) affect the relationship between the SCR's forward voltage and forward current. The $I_G = 0$ curve shows the relationship between V_F and I_F when the gate current is zero and is therefore simply a more detailed representation of the forward characteristics shown in Figure 8-4. The I_{G1} curve is plotted for a specific but relatively low value of gate current. Notice that this curve has the same general shape as the $I_G = 0$ curve, but the forward breakover point occurs sooner (at a lower V_F value). The I_{G2} curve is plotted for a slightly higher gate current and also has the same general shape as the other two curves. However, the breakover point occurs even sooner at this higher value of gate current.

The curves in Figure 8-5 show that the SCR's forward breakover voltage decreases as the gate current increases. In fact, the gate current could be increased to a point where the breakover voltage would be so low that the device would have characteristics that closely resemble those of an ordinary PN junction diode. The ability of the gate to control the point where breakover occurs is an advantage in many types of electronic circuits.

The curves in Figure 8-5 reveal the SCR's most important electrical characteristics. Basically these curves show that for any given gate current, a specific forward breakover voltage must be reached before the SCR can turn on. However, the curves also show that for any given forward voltage across the SCR, a specific value of gate current must be reached before the device can be turned on. Therefore, an SCR can be turned on only when it is subjected to the proper combination of gate current and forward voltage values.

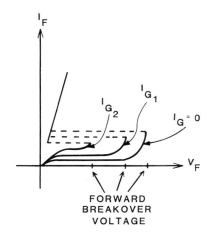

Figure 8-5 Forward characteristics of the SCR for different values of gate current.

Practical Applications

In normal operation the SCR is subjected to forward voltages which are below its breakover voltage and the SCR is made to turn on by the application of a suitable gate current. This gate current is usually made high enough to insure that the SCR is switched to the on state at the proper time. Furthermore, the gate current is usually applied for just an instant in the form of a current pulse. A constant gate current is not required to trigger the SCR and would only cause more power to be dissipated within the device. Once the SCR turns on, it can be turned off only by reducing its forward current below its respective holding current value.

The SCR is primarily used to control the application of DC or AC power to various types of loads. It can be used to open or close a circuit or it can be used to vary the amount of power applied to a load. A very low gate current signal can control a very large load current. The SCR in Figure 8-3 is used as a switch to apply DC power to the load resistor (R_2) but in this basic circuit there is no effective means of turning off the SCR and thus remove power from the load. However, this problem can be easily solved by simply connecting a switch across the SCR. This switch can be momentarily closed to short out the SCR and reduce its anode-to-cathode voltage to zero. This will reduce the SCR's forward current below the holding value and cause the SCR to turn off.

A more practical SCR circuit is shown in Figure 8-6. With this circuit, mechanical switches have been completely eliminated. In this high speed circuit SCR_1 is used to control the DC power applied to load resistor R_L and SCR_2 along with a capacitor (C) and a resistor (R_1) are used to turn off the circuit. When a momentary gate current flows through SCR_1, it turns on and allows a DC voltage to be applied to R_L. This effectively grounds the left side of capacitor C and allows it to charge through resistor R_1.

This, in turn, causes the right hand plate of the capacitor to become positive with respect to the left hand plate. When a momentary gate current pulse is applied to SCR$_2$, it turns on and the right hand plate of the capacitor is grounded thus placing the capacitor across SCR$_1$. The voltage across the capacitor now causes SCR$_1$ to be reverse-biased. This reverse voltage causes the forward current through SCR$_1$ to drop below its holding value, thus, causing SCR$_1$ to turn off and remove power from R$_L$. Therefore a momentary gate current through SCR$_1$ will turn on the circuit and a momentary gate current through SCR$_2$ will turn off the circuit.

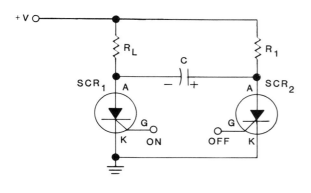

Figure 8-6 A practical DC SCR switch control.

When using SCR's in DC switching circuits it is often necessary to use additional components (not always additional SCRs) to provide a means of turning off the circuits. The previous example shows only one of the ways this can be accomplished.

An SCR may also be used to control the application of AC power to a load. However, when used in AC circuits, the device is capable of operating on only one alternation of each AC input cycle. A simple AC switch circuit is shown in Figure 8-7. The SCR can conduct only on those alternations which make its anode positive with respect to its cathode. Furthermore, switch S must be closed so that gate current will flow through the SCR and allow it to conduct. Resistor R$_1$ limits the peak value of this gate current and diode D$_1$ prevents a reverse voltage from appearing between the SCR's gate and cathode during the reverse portion of each cycle.

By closing switch S the SCR is allowed to conduct on one alternation of each input cycle and applies this portion of the AC voltage to load resistor R$_L$. When switch S is opened, the SCR will turn off within one-half cycle of the AC signal. In other words, when no gate current can flow through the SCR, the device will turn off as soon as the input AC voltage drops to zero. The SCR will remain off until the switch is again closed.

The AC switch circuit in Figure 8-7 is actually less complicated than the DC circuit in Figure 8-6. In the AC circuit no additional components are required to turn off the SCR since this occurs automatically when the AC input voltage drops to zero. Instead, the additional components are used to insure that the SCR turns on during the proper portion of each cycle. It is also important to note that this switch circuit can deliver only half of the available AC power to the load since the SCR can conduct only on one alternation of each AC cycle. However, other circuit arrangements can be used which will allow full AC power to be applied to the load. This is often accomplished by first rectifying the AC input signal so that both alternations of each input cycle will properly bias the SCR into conduction. In other words the positive and negative alternations of each cycle are converted to all positive or all negative alternations. The pulses are then considered to be pulsating DC current pulses.

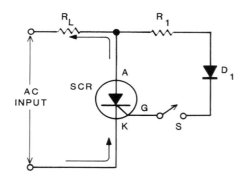

Figure 8-7 Simple AC SCR switch.

The circuits in Figures 8-6 and 8-7 are used to apply or remove electrical power and to take the place of a mechanical switch. However, when compared to mechanical switches or relays, these SCR circuits have many advantages. They do not wear out like mechanical devices and they do not have contacts which can bounce or stick and

cause intermittent operation. The SCR circuits are also more reliable than mechanical devices where large amounts of power must be controlled. SCR circuits may in turn be controlled mechanically or electrically. In either case, it is only necessary to control the SCR's very small gate current. If this is done mechanically, a relatively inexpensive switch that has low current and voltage rating can be used.

SCR's may also be used to vary the amount of power applied to a load instead of just simply switching the power on or off. In fact, they are widely used in power control applications where the source of power is 60 Hz AC. One of the most basic AC power control circuits that utilizes an SCR is shown in Figure 8-8. This circuit is commonly referred to as a half-wave phase control circuit. It uses just one SCR and it is capable of controlling the AC power that is applied to load resistor R_L.

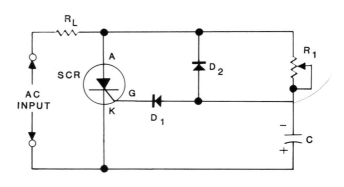

Figure 8-8 Half-wave phase control SCR circuit.

The AC voltage applied to the control circuit in Figure 8-8 is the standard 120 volts, 60 Hz sine wave. Two complete cycles of this waveform are shown in Figure 8-9A. On each negative alternation of the input voltage, capacitor C charges through the forward biased diode D_2 and the SCR is biased in the reverse direction so that it cannot conduct. Also, on each negative alternation, diode D_1 is reverse-biased and will not allow gate current to flow through the SCR.

During each positive alternation the SCR is forward-biased so that it can conduct forward current through R_L if its gate current is high enough to turn it on. However, the SCR's gate current is controlled by resistor R_1 and capacitor C. The capacitor will discharge and then recharge through R_1 during the positive alternations since D_2 is reverse-biased during these portions of the AC cycle. Furthermore, the rate at which the capacitor charges can be controlled by adjusting the resistance of R_1.

If the resistance of R_1 is set to zero, capacitor C will charge almost immediately and the voltage across capacitor C (which is connected to the SCR's gate through D_1) will quickly rise to a level which will cause the SCR to turn on. In fact, when R_1 is equal to zero the SCR will turn on almost at the beginning of each positive alternation and apply power to R_L for the entire alternation. Once the SCR turns on it will not turn off until the AC input voltage drops to zero. Therefore, when R_1 is equal to zero the SCR acts like an ordinary junction diode and conducts during each positive alternation of

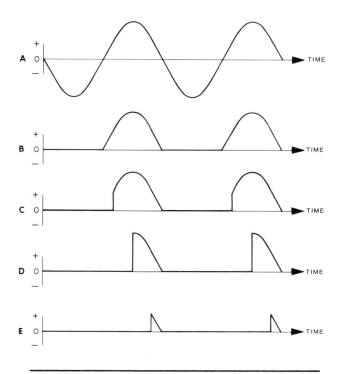

Figure 8-9 Waveforms for half-wave phase control SCR circuit.

the input AC voltage. This causes the voltage across R_L to appear as shown in Figure 8-9B.

When the resistance of R_1 is increased slightly, capacitor C cannot discharge and then recharge as quickly during each positive alternation. This means that it will take slightly longer for the voltage across capacitor C to rise to a level which will cause the SCR to turn on. However, when the SCR does turn on, it remains on for the remainder of each positive alternation. This means that the SCR turns on shortly after each positive alternation has started and not at the beginning of each alternation as it did before. This causes the voltage across R_L to appear as shown in Figure 8-9C.

The resistance of R_1 can be further increased to further extend the time required for the SCR to turn on during each positive alternation. Figure 8-9D shows the voltage across R_L when R_1 is increased to the point where the SCR conducts for only half of each positive alternation and Figure 8-9E shows the voltage across R_L when R_1 is made even larger. If the value of R_1 is increased further, the SCR will not conduct at all and no power will be applied to R_L.

As shown in Figure 8-9, the half-wave phase control circuit is capable of controlling the amount of power applied to the load. With this circuit, the power applied to the load can be varied from zero to approximately 50 percent of the input AC power.

It is also possible to use an SCR to control current during both alternations of an AC signal, however it is necessary to first rectify the AC signal so that both alternations of each cycle are made to flow in the same direction (known as full wave rectification) before being applied to the SCR. Another method of achieving full control of an AC signal is to use two SCRs that are connected in parallel but in opposite directions. This arrangement allows the SCRs to conduct alternately. Either of these two methods will allow full control of the input AC signal so that the power applied to the load can be varied from zero to the full value of the AC input.

1. An SCR has _____ semiconductor layers.

2. The SCR has three leads which are referred to as the:

    _____

    _____

    _____

3. An SCR is used basically as an electronic:

    _____

4. The SCR can be turned on or off and therefore has two stable states.

    A. True
    B. False

5. The forward voltage required to turn on an SCR is referred to as the SCR's:

    _____ _____ _____.

6. The minimum value of forward current required to keep an SCR in the on state is referred to as the SCR's:

    _____ _____

7. The reverse voltage that will cause the SCR's reverse current to change from an extremely small leakage value to a relatively high value is called the SCR's:

    _____ _____ _____

8. An SCR's forward breakover voltage will decrease when its gate current is:

    _____

9. Under normal conditions an SCR is biased so that it can be turned on with the application of a momentary _____ current.

10. An SCR may be used to control the amount of AC power applied to a load.

    A. True
    B. False

Bidirectional Triode Thyristors

The silicon controlled rectifier (SCR) previously described is capable of controlling current which is flowing in one direction and is therefore a unidirectional device. The SCR is used in many applications which involve the control of direct currents as well as alternating currents. Unfortunately when used in AC applications, a single SCR is capable of operating on just one alternation of each AC input cycle. In order to achieve full control of each AC input cycle, it is necessary to use two SCR's in parallel or it is necessary to convert each entire AC cycle into a pulsating DC signal before it is applied to a single SCR.

In applications where it is necessary to achieve full control of an AC signal, it is often much easier to use a device known as a bidirectional triode thyristor. This device is more commonly referred to as a triac. The triac has basically the same switching characteristics as an SCR, however, it exhibits these same characteristics in both directions. This makes the triac equivalent to two SCRs which are in parallel but are connected in opposite directions.

We will now briefly examine the triac's basic construction and operation and then we will consider its important characteristics and applications.

Basic Construction and Operation

A simplified diagram of a triac is shown in Figure 8-10A. Notice that the device has three leads which are designated as main terminal 1, main terminal 2 and the gate. Main terminal 1 and main terminal 2 are each connected to a PN junction at opposite ends of the device. The gate is also connected to a PN junction which is at the same end as terminal 1. If you examine the entire structure closely you will see that from terminal 1 to terminal 2 you can pass through an NPNP series of layers or a PNPN series of layers. In other words the triac is effectively a four-layer NPNP device in parallel with a four-layer PNPN device. These NPNP and PNPN devices are often compared to two SCRs which are connected in parallel but in opposite directions. The equivalent SCR arrangement is shown in Figure 8-10B.

The triac's gate region is more complex and a detailed analysis of its operation will not be considered at this time. However, the gate is basically capable of directly or remotely triggering either of the equivalent SCRs into conduction. Notice that both of the equivalent SCR gates are tied together in Figure 8-10B to show the equivalent relationship.

The circuit in Figure 8-10B is not in all ways equivalent to the triac. This circuit is used simply to explain the basic concept involved. The primary difference is that the two equivalent SCRs would actually require different gating circuits to trigger them into conduction but the triac is designed to respond to the currents that flow through its single gate terminal.

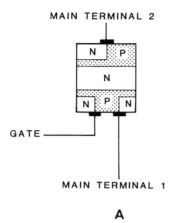

MAIN TERMINAL 2

N P

N

N P N

GATE

MAIN TERMINAL 1

A

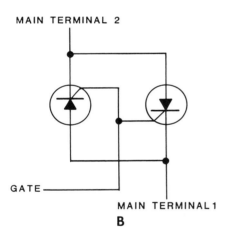

MAIN TERMINAL 2

GATE

MAIN TERMINAL 1

B

Figure 8-10 The triac and its equivalent circuit.

Unlike an SCR which can control currents flowing in only one direction, the triac can control currents flowing in either direction. The triac is therefore widely used to control the application of AC power to various types of loads or circuits. The conditions required to turn a triac on or off in either di-

rection are similar to the conditions required to control the SCR. Both devices can be triggered to the on state by a gate current and they can be turned off by reducing their operating anode currents below their respective holding values. In the case of an SCR, current must flow in the forward direction from cathode to anode. However, the triac is designed to conduct both forward and reverse currents through its main terminals.

The schematic symbol that is commonly used to represent the triac is shown in Figure 8-11. Notice that the symbol consists of two parallel diodes connected in opposite directions with a single gate lead attached. The device is usually placed within a circle as shown and its main terminals are sometimes identified as MT_1 and MT_2 as indicated.

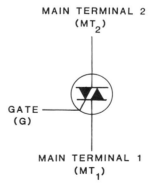

MAIN TERMINAL 2
(MT_2)

GATE
(G)

MAIN TERMINAL 1
(MT_1)

Figure 8-11 Schematic symbol of a triac.

V-I Characteristics

The voltage-current (V-I) characteristic curve for a typical triac is shown in Figure 8-12. This curve shows the relationship between the current flowing through its main terminals in each direction (designated as $+I_T$ and $-I_T$) and the voltage applied across its main terminals in each direction (identified as $+V$ and $-V$). Furthermore, this curve was plotted with no gate current flowing through the triac and main terminal 1 was used as the reference point for all voltage and current values.

Figure 8-12 shows that when main terminal 2 (MT_2) is positive with respect to MT_1 (or when the applied voltage is equal to $+V$), the current through the device ($+I_T$) remains at a low leakage value until $+V$ rises above the breakover voltage ($+V_{BO}$) of the device. At this time the triac switches from the off state to the on state and $+I_T$ is essentially limited by the external resistance of the circuit. The triac must remain in the on state until $+I_T$ drops below a specified holding current (I_H) as shown. This is exactly what happens when an SCR is subjected to a forward voltage that exceeds its respective forward breakover voltage.

When MT_2 is negative with respect to MT_1, the triac exhibits the same basic V-I characteristics since the current through the device ($-I_T$) remains at a low leakage value until $-V$ rises above the breakover voltage ($-V_{BO}$) of the device. At this time the triac switches from the off state to the on state and remains on until $-I_T$ drops below I_H.

The V-I curve in Figure 8-12 therefore shows that the triac exhibits the characteristics of an SCR in either direction. However, this curve does not show how the triac's gate is used to control its operation. Like the SCR previously described, the triac's breakover voltage (in either direction) can be varied by controlling the amount of gate current flowing through the device. When the gate current is increased, the breakover voltage is lowered. However, there is still one very important difference between the SCR's gating characteristics and those of a

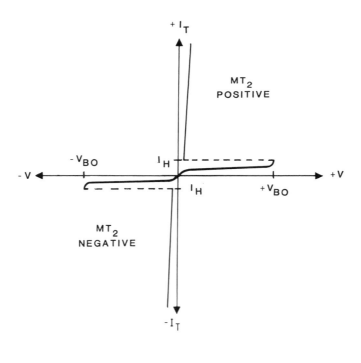

Figure 8-12 V-I characteristics of a typical triac.

triac. The SCR always requires a positive gate voltage, but the triac will respond to either a positive or negative gate voltage. In other words, the triac's breakover voltage (in either direction) can be lowered by making its gate more positive or more negative with respect to MT_1, which is used as the reference terminal. This positive or negative gate voltage correspondingly produces a gate current that flows out of or into the gate lead. These currents in turn regulate the point at which the device turns on.

Like the SCR, the triac is normally subjected to operating voltages that are below its breakover voltage (in either direction). The device is turned on by subjecting it to a sufficiently high gate current which flows into or out of its gate lead. The device is turned off by simply reducing its operating current ($+I_T$ or $-I_T$) below its respective I_H value. The triac is most sensitive when it is subjected to $+V$ and $+I_T$ values along with a positive gate voltage. Under these conditions the device requires the least gate current to turn on for any given $+V$ value. Other combinations of operating voltages and currents result in a loss of sensitivity. To help the circuit designer determine the conditions necessary to turn on a specific triac, manufacturers of these components usually specify minimum or typical values of gate current (in each direction) required to turn on the device. These values are given for a specified operating voltage which is applied in first one direction and then the other ($+V$ and $-V$). With this information the circuit designer can insure that suffi-cient gate current is used to turn on the triac at the proper time. As with the SCR circuits, this gate current need only be applied momentarily to cause the triac to change states.

Applications

Since the triac conducts in either direction it is ideally suited for applications where AC power must be controlled. The device can be used as an AC switch or it can be used to actually control the amount of AC power applied to a load. A typical example of its use as an AC switch is shown in Figure 8-13. This circuit will apply the full input

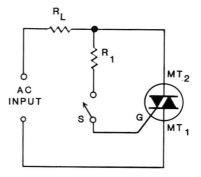

Figure 8-13 Simple AC switch.

voltage across load resistor R_L or completely remove it when switch S is closed or opened respectively. When switch S is open the triac cannot conduct on either the positive or the negative alternations of each AC input cycle. This is because the input voltage does not exceed the triac's breakover voltage in either direction. However, when switch S is closed, resistor R_1 allows enough gate current to flow through the triac on each alternation to insure that the device turns on. The triac therefore applies all of the available input power to the load while a comparable SCR circuit (refer to Figure 8-7) is capable of supplying only half of the input power to the load. The advantage of this circuit is that the small gate current can control a high load current.

A typical triac circuit which can be used to vary the amount of AC power applied to a load is shown in Figure 8-14. This circuit is generally referred to as a full wave phase control circuit and it operates in a manner similar to the SCR circuit shown in Figure 8-8. The primary difference is that the triac is triggered into conduction on both the positive and negative alternations of each AC input cycle, while the SCR in Figure 8-8 conducts only on positive alternations. Also, a special triggering device is generally used to insure that the triac turns on at the proper time.

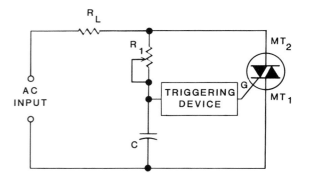

Figure 8-14 Full-wave phase control circuit.

Capacitor C charges through R_1 in first one direction and then the other as the positive and negative alternations of the AC input signal occur. During each alternation, the triac is turned on when the voltage across capacitor C rises to the required level. However, this voltage is not applied directly to the triac's gate and MT_1 leads. Instead, it is applied through a special triggering device which has bidirectional switching characteristics. The triggering device can be any component which will turn on or conduct when subjected to a specific voltage level and turn off when the voltage is reduced to a lower level. One of the most widely used triggering devices will be described in detail later in this unit. However, at the present time we will just assume that the device is a solid-state component with the switching characteristics just described.

During each alternation, the voltage across capacitor C rises to a level which turns the triggering device on. This causes current to momentarily flow through the triac's gate lead and switches it to the on state. The gate current only flows for a moment since the capacitor discharges through the triac and loses its accumulated voltage which causes the triggering device to turn off. How soon the triac turns on during each alternation is determined by the value of R_1. When the resistance of R_1 is reduced to zero, the triac is triggered immediately at the beginning of each alternation and full AC power is applied to load resistor R_L. As the resistance of R1 is increased, triggering occurs later during each alternation and the average power applied to the load is reduced. The voltage waveforms shown in Figure 8-9 can also represent the voltage across R_L in the triac circuit if the negative alternations (which are identical but complementary to the positive alternations) are inserted.

A triggering device is required because the triac is not equally sensitive to gate currents flowing in opposite directions as explained earlier. The triggering device helps to compensate for the triac's non-symmetrical or non-uniform triggering characteristics. The voltage required to turn on the triggering device is identical in both directions and the device is designed to be as insensitive to temperature changes as possible. The triggering device works in conjunction with resistor R_1 and capacitor C to produce consistently accurate gate current pulses that are high enough to turn the triac on at the prop-

er time in either direction. These gate current pulses can be very short in duration (several microseconds is usually sufficient) and still trigger the triac.

Although the triac has the ability to control current in either direction and respond to gate currents flowing in either direction, the device does have certain disadvantages when compared to an SCR. In general, triacs have lower current ratings than SCRs and cannot compete with the SCRs in applications where extremely large currents must be controlled. Triacs are available that can handle currents (usually measured in rms values) as high as 25 amperes. By comparison, SCRs can be readily obtained with current ratings (usually expressed as average values for a half cycle) as high as 700 to 800 amperes, and some are rated even higher. Also, both devices can have peak or surge current ratings that are much higher than their respective rms or average ratings.

Triacs often have difficulty in switching the power applied to inductive loads. This problem also occurs with SCRs but to a lesser degree. When triacs are used to control the power applied to inductive loads such as motor windings or heater coils, it is always necessary to use additional components to improve their operation. Also, triacs are designed for low frequency (50 to 400 Hz) applications while SCRs can be used at frequencies up to 30 kHz. Therefore, in certain applications where full control of an AC signal is required, SRCs may operate more efficiently than triacs while in other applications the exact opposite may be true.

11. The triac is often compared to two _____ which are connected in parallel but in opposite directions.

12. The triac's gate is capable of turning on the device so that it can conduct current in either direction through its MT_1 and MT_2 leads.

    A. True
    B. False

13. When the triac turns on in either direction, it can be turned off only by reducing its main operating current ($+I_T$ or $-I_T$) below a specified _____ current value.

14. The triac's breakdown voltage in either direction is lowered when its gate current is _____.

15. When a triac is subjected to an operating voltage that is less than its breakdown voltage rating (in either direction) the device can be turned on by a gate current that is flowing in either direction.

    A. True
    B. False

16. A triac exhibits the same sensitivity to gate currents flowing in either direction.

    A. True
    B. False

17. When a triac is used to control the amount of AC power applied to a load, a special _____ device is used in conjunction with the triac to compensate for the triac's non-symmetrical triggering characteristics.

18. The triac in Figure 8-14 will apply full AC power to R_L when R_1 is equal to _____.

19. Triacs should always be used instead of SCRs in application where full control of AC power is required.

    A. True
    B. False

Bidirectional Trigger Diodes

As explained earlier, a triggering device is used in conjunction with the triac because the triac does not have symmetrical triggering characteristics. Various types of triggering devices can be used with the triac, but one of the most popular devices is known as a bidirectional trigger diode, commonly referred to as a diac.

We will now briefly examine the construction and operation of this important triggering device. Then we will see how it is used in conjunction with the triac.

Basic Construction and Operation

The diac is constructed in much the same way as a bipolar transistor. The device has three alternately doped semiconductor layers as shown in Figure 8-15A. However, it differs from the bipolar transistor because the doping concentrations around both junctions are equal and leads are attached only to the outer layers. No electrical connections are made to its middle region. Since the diac has only two leads, it is often packaged in a metal or plastic case which has axial leads. Therefore the device often resembles an ordinary PN junction diode in appearance. However, the device is sometimes packaged like a conventional bipolar transistor but with only two leads.

Since both of its junctions are equally doped, the diac has the same effect on currents flowing in either direction through its leads. In either direction, one junction will always be forward-biased while the other is reverse-biased. In each case the reverse-biased junction primarily controls the current flowing through the diac and the device operates as if it contains two PN junction diodes that are in series but are connected back-to-back. See Figure 8-15C.

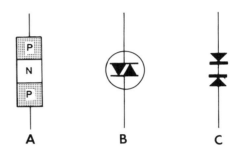

Figure 8-15 The diac, its schematic symbol and equivalent circuit.

The diac remains in an off state (conducts only a small leakage current) in either direction until the applied voltage in either direction is high enough to cause its respective reverse-biased junction to break down. When this happens, the device turns on and current suddenly rises to a value which is essentially limited by the resistance in series with the device. The diac therefore functions as a bidirectional switch which will turn on whenever its breakdown voltage (in either direction) is exceeded.

A schematic symbol that is commonly used to represent the diac is shown in Figure 8-15B. Notice that it is similar to the triac symbol, except that a gate lead is not required.

V-I Characteristics

The voltage-current (V-I) characteristic curve for a typical diac is shown in Figure 8-16. This curve shows the relationship between the current flowing through the device in either direction ($+I$ and $-I$) and the corresponding voltage across the device in each direction ($+V$ and $-V$).

Figure 8-16 shows that the current through the diac remains at a low value until the voltage across the device increases to a point where the device breaks down in either direction. These voltages required to breakdown in the diac are generally referred to as the diac's breakover voltages and are designated as $+V_{BO}$ and $-V_{BO}$ as shown in Figure 8-16. The $+V_{BO}$ and $-V_{BO}$ values for a typical device will usually be between 28 and 36 volts. Notice that once the $+V_{BO}$ and $-V_{BO}$ values are reached, the current through the diac increases rapidly but the voltage across the device decreases. In other words when breakover occurs, the resistance of the diac decreases rapidly as current increases and the net result is a decrease in voltage across the device. Therefore, once the breakover point is reached the diac exhibits a negative resistance characteristic.

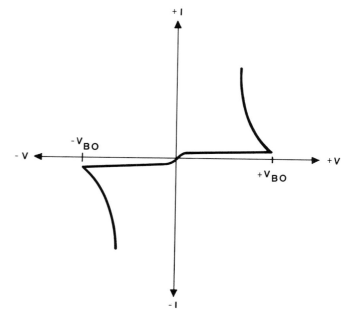

Figure 8-16 V-I Characteristics of a typical diac.

The diac is considered to be in the off state until the breakover voltage ($+V_{BO}$ or $-V_{BO}$) is reached. When $+V_{BO}$ or $-V_{BO}$ is reached, the device switches to the on state and it is capable of conducting relatively high currents.

Applications

The diac is most commonly used as a triggering device for a triac which is, in turn, used to control the amount of AC power applied to a load. This application is shown in Figure 8-17. The circuit shown in this figure is equivalent to the full-wave phase control circuit shown in Figure 8-14. The only difference is that a diac is specially used as a triggering device. As explained earlier, the triggering device turns on when the voltage across C_1 rises to the required value. In this case the voltage across C_1 must rise to either $+V_{BO}$ or $-V_{BO}$ before the diac can switch from the off state to the on state. Each time the diac turns on, it allows current to momentarily flow through the triac's gate and turn it on. The gate current flows only momentarily because it is provided by capacitor C_1 which quickly discharges through the triac's MT_1 and gate terminals and the diac. The peak value and duration of the current pulse applied to the gate are determined by the value of C_1, the resistance of the diac (which changes with current), and the resistance between the triac's gate and MT_1 terminals. To help designers relate these various factors so that the proper gate triggering will be obtained, manufacturers of diacs and triacs generally provide applicable charts or curves. The relationship between triggering capacitance (the value of C_1) and peak current for the diac is usually shown in this manner and various curves are used to show the minimum and maximum limits on the peak amplitude and duration of the gate current pulses that can be applied to a triac.

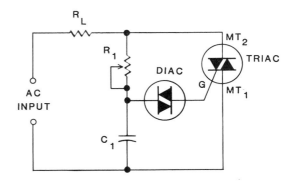

Figure 8-17 Full-wave phase control circuit using a diac and a triac.

The diac is used in conjunction with the triac to provide full-wave control of AC signals. These devices are commonly used together to control the speed and direction of electric motors or control the temperature of heating elements. They are used in air conditioning, heating, and ventilation systems and they can even be used in electronic garage-door systems. They could also be used to switch or control the amount of power applied to individual lamps or complete lighting systems. For example, they can be used to flash a lamp on and off as a warning signal, or to control traffic-signal lights which are placed at roadway intersections to regulate traffic flow, or to provide light dimming in theaters. Diac-triac combination circuits can also be used to control power in three-phase systems by using multiple triac circuit arrangements.

Although the diac is widely used to trigger the triac in power control applications, it is important to realize that other triggering devices are also available. Most of these triggering devices have bidirectional switching characteristics that are closely related to those of a diac, but each device is constructed differently and each is represented by a different schematic symbol. The diac just described serves as a typical example of a modern solid-state triggering device.

20. The diac has three semiconductor layers but only _____ terminals.

21. The diac turns on when its _____ voltage is reached in either direction.

22. When the diac turns on in either direction, the device exhibits a _____ resistance.

23. When the diac is turned on, the voltage across the device decreases as the current through the device _____.

24. The diac is used in conjunction with a _____ to control the AC power applied to a load.

25. In Figure 8-17 the diac turns on when the voltage across _____ rises to the diac's $+V_{BO}$ or $-V_{BO}$ values.

26. The diac is the only triggering device which can be used to control the triac.

    A. True
    B. False

Unijunction Transistors

We will now examine another important member of the thyristor family which is known as a unijunction transistor or simply a UJT. The UJT has physical and electrical characteristics that are quite different from those of the diacs and triacs previously described. The UJT is actually a special type of transistor which is used as an electronic switch and not as an amplifying device. Furthermore, the UJT can be used to generate repetitive waveforms which can be used to perform many useful functions in electronic circuits.

In addition to the basic unijunction transistor, a special type of UJT is also used. This device is commonly referred to as a programmed unijunction transistor or simply a PUT. The PUT operates in basically the same manner as the UJT but its electrical characteristics can be made to vary over a wide range.

We will now examine the physical and electrical characteristics of the basic UJT and see how it is used as an electronic switch and waveform generator. Then we will briefly examine the special PUT and consider its advantages when compared to the UJT.

Thyristors can be packaged in various ways as shown in Figure 8-18. The large unit on the left is a stud-mounted SCR which can handle currents as high as 110 amperes and the three components next to it are SCRs with lower current ratings. The component at the extreme right is a plastic encapsulated SCR with a current rating of 4.5 amperes.

The large stud-mounted component fourth from the right is a triac with a current rating of 50 amperes.

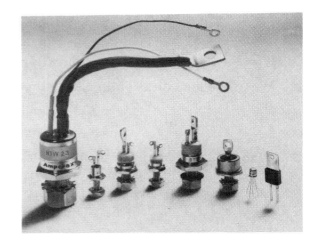

Figure 8-18 Typical examples of the thyristor family (courtesy of Amperex Electronic Corp.)

Basic Construction

As its name implies, the unijunction transistor (UJT) has only one semiconductor junction. The device basically consists of a block or bar of N-type semiconductor material which has a small pellet of P-type material fused into its structure as shown in Figure 8-19A. A lead is attached to each end of the N-type bar as shown and these two leads are referred to as base 1 and base 2. Another lead is attached to the P-type pellet and referred to as the emitter. These leads are commonly designated as B_1, B_2, and E as shown.

The N-type bar is lightly doped and therefore has few majority carriers to support current flow. This means that the resistance between the base 1 and base 2 leads is quite high. Most UJTs exhibit an interbase resistance (between B_1 and B_2) that is approximately 5 k ohms to 9 k ohms.

The area in which the P-type pellet and N-type bar meet to form the PN junction, has characteristics similar to those of a PN junction diode. Therefore, we can represent the UJT with the equivalent circuit shown in Figure 8-19B. This equivalent circuit consists of two resistors (R_{B_1} and R_{B_2}) and a diode which has its cathode end connected between the resistors. The operation of the UJT is much easier to understand when this equivalent circuit is used in place of the UJT's basic structure as shown in Figure 8-19A.

The schematic symbol that is commonly used to represent the UJT is shown in Figure 8-19C. The arrow on the emitter (E) lead points inward to show that the UJT has a P-type emitter. The three leads may not always be identified as shown.

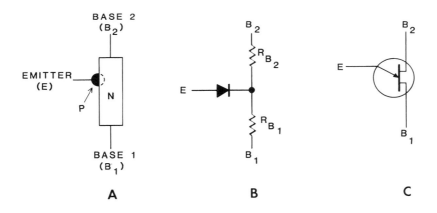

Figure 8-19 The basic UJT with its equivalent circuit and its schematic symbol.

Operation

The operation of an ordinary UJT is more apparent when its equivalent circuit is analyzed. Figure 8-20 shows how the equivalent UJT circuit is biased under normal conditions. Notice that an external voltage source (V_{BB}) is connected across the B_1 and B_2 terminals so that B_2 is positive with respect to B_1. Another external voltage source (V_S) is connected across the E and B_1 terminals so that E is positive with respect to B_1. A resistor is placed between the positive side of V_S and terminal E. This resistor is used to limit the current through E to a safe level.

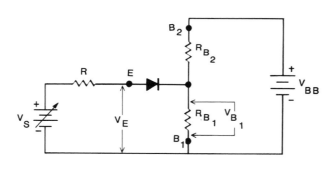

Figure 8-20 A properly biased UJT.

If voltage VS is not high enough to cause the diode in the equivalent circuit to be forward biased, the two resistors within the equivalent circuit (R_{B_1} and R_{B_2}) will allow only a small current to flow between termi-

nals B_1 and B_2. This current could be easily calculated according to Ohm's Law by dividing the total resistance between terminals B_1 and B_2 ($R_{B_1} + R_{B_2}$) into voltage V_{BB}. This means that voltage V_{BB} is distributed across R_{B_1} and R_{B_2}. The ratio of the voltage across R_{B_1} (designated as V_{B_1}) to the source voltage (V_{BB}) is known as the intrinsic standoff ratio and is represented by the Greek letter (eta). This relationship can be expressed mathematically as:

$$\text{intrinsic standoff ratio } (\eta) = \frac{V_{B_1}}{V_{BB}}$$

Since the voltage dropped across R_{B_1} and R_{B_2}, is proportional to their resistance values, the intrinsic standoff ratio is also equal to the ratio of R_{B_1} to the total resistance between terminals B_1 and B_2 ($R_{B_1} + R_{B_2}$). This can be expressed mathematially as:

$$\eta = \frac{R_{B_1}}{R_{B_1} + R_{B_2}}$$

This last equation shows that the intrinsic standoff ratio is basically determined by the two internal resistances. Its value is determined by the physical construction of the device and it cannot be controlled by varying V_{BB} or V_S. The intrinsic standoff ratio is specified for each type of UJT that is made and typical values will range from approximately 0.5 to 0.8. When the intrinsic stand-off ratio for a particular device is known, the voltage across R_{B_1} (V_{B_1}) can be determined for any value of applied voltage (V_{BB}). This calculation can be made by simply transposing the first equation given to obtain:

$$V_{B_1} = \eta \, V_{BB}$$

This equation simply states that the voltage across R_{B_1} (V_{B_1}) is equal to the intrinsic standoff ratio times V_{BB}. For example, if the UJT has an intrinsic standoff ratio of 0.5 and it is subjected to a V_{BB} of 20 volts, the voltage across R_{B_1} would be equal to 0.5 times 20 or 10 volts.

The UJT functions in the manner just described as long as V_S is not high enough to forward-bias the diode and cause it to conduct. However, the UJT will exhibit different characteristics when V_S is high enough to forward bias the diode. In order for this to happen, V_S must be increased until the voltage that appears between the emitter (E) and B_1 terminals (designated V_E) is higher than the voltage across R_{B_1} and the voltage required to turn on the diode which is approximately 0.7 volts. In other words, the voltage across R_{B_1} (V_{B_1}) causes the diode to be reverse-biased and this voltage must be completely cancelled by the opposing input voltage (V_E). Then V_E must be raised an additional 0.7 volts above V_{B_1} so that the diode will become forward-biased by a voltage that is high enough to cause it to conduct.

The value of V_E required to turn on the diode is called the peak voltage and is usually designated as V_P. The value of V_P is determined by the source voltage V_{BB}, the intrinsic standoff ratio, and the voltage required to turn on the diode. This relationship can be shown mathematically as:

$$\text{peak voltage } (V_P) = \eta \, V_{BB} + V_F$$

This equation simply states that V_P is equal to the product of the intrinsic standoff ratio and source voltage plus the voltage required to turn the diode on (V_F). Manufacturers usually do not indicate the value of V_P for a particular UJT since it varies with V_{BB}. For example, if a UJT has an (η) of 0.6 and a V_{BB} of 10 volts, its peak voltage will be:

$$V_P = (0.6)\,(10) + 0.7$$
$$V_P = 6 + 0.7$$
$$V_P = 6.7 \text{ volts}$$

This means that the diode will turn on when input voltage V_E reaches 6.7 volts.

Until the V_P value is reached, the diode conducts only a very small reverse leakage current which flows through the emitter lead. However, when the V_P value is reached, the diode turns on and allows current to flow in the forward direction through its PN junction and through the emitter lead. This current (often designated as I_E) results because the diode turns on, but an additional action also takes place. When I_E begins to flow in the forward direction, many charge carriers (holes) are injected into the lightly doped N-type bar and are swept toward terminal B_1 which is negative with respect to the emitter. These charge carriers increase the conductivity of the bar between the emitter (E) and B_1 which of course means that the resistance of R_{B_1} is reduced. This lower resistance causes I_E to increase even further and more charge carriers are injected into the bar which in turn lowers the value of R_{B_1} even further. This in turn causes I_E to increase even further. This action is cumulative and it starts to occur when the V_P value is reached. At this time the UJT is said to be turned on.

If the source voltage (V_S) is increased further, the cumulative action just described becomes even more apparent. As V_S increases, I_E increases rapidly due to the decrease in the resistance of R_{B_1}. Furthermore, this decrease in the value of R_{B_1} causes the voltage V_E, which appears between the E and B_1 terminals, to decrease even though I_E increases in value. The UJT therefore exhibits a negative resistance characteristic after it turns on.

If V_S is increased even further, I_E would continue to increase, however a point would eventually be reached where V_E would stop decreasing and actually begin to rise slightly. This point marks the end of the negative resistance region. Beyond this point an increase in I_E is accompanied by a slight increase in V_E.

So far we have considered only the action that takes place between terminals E and B_1

when the UJT turns on. We have seen that R_{B_1} decreases, however it is important to note that R_{B_2} also decreases a certain amount since some of the injected charge carriers (holes) enter that portion of the N-type bar. However, the decrease in the value of R_{B_2} is small when compared to the decrease in R_{B_1}. This means that the total resistance of the bar ($R_{B_1} + R_{B_2}$) is reduced once the UJT turns on and this results in an increase in current through these two resistors and voltage source V_{BB}. However, R_{B_2} does not decrease by much and therefore does not allow the current to rise to a high value.

The most important action takes place between terminals E and B_1. It is this portion of the UJT that provides the most useful characteristics. The two most important features are the ability to turn on at a specific V_P value and the negative resistance characteristic that occurs for a certain period of time after the device turns on.

V-I Characteristics

The action that takes place between the E and B_1 terminals of an ordinary UJT is shown graphically in Figure 8-21. The curve in this figure shows the relationship between the current flowing through the emitter of a typical UJT (designated as I_E) and the voltage appearing across its E and B_1 terminals (designated as V_E). Such a curve could be plotted by using a circuit like the one shown in Figure 8-20. The V_E and I_E values could be observed while V_S is varied over a wide range.

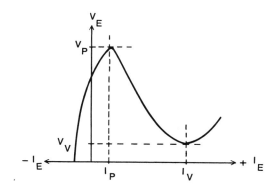

Figure 8-21 V-I characteristic between E and B1 terminals of a typical UJT.

Figure 8-21 shows that when V_E equals zero a small negative current $(-I_E)$ flows through the emitter lead. This is a small leakage current which flows from left to right through the diode (refer to Figure 8-20) because of the relatively large voltage across R_{B_1} (V_{B_1}). As V_E increases, it opposes V_{B_1}, and the leakage current decreases. When V_E is equal to V_{B_1} the current is reduced to zero and any further increase in V_E results in a positive current $(+I_E)$ which flows from right to left through the diode. When V_E reaches the peak voltage (V_P) value, the UJT is considered to be in the on state. The current that flows at this point is called the peak current and is designated as I_P as shown in Figure 8-21.

Beyond the V_P point, V_E decreases as $+I_E$ increases, thus, giving the device a negative resistance characteristic. This negative resistance continues until V_E starts to increase

again. The point where V_E reaches its minimum value and starts to increase is called the valley voltage and is designated as V_V. The current flowing at this time is referred to as the valley current or I_V. Beyond the V_V point, V_E increases slightly as $+I_E$ increases and the UJT no longer exhibits a negative resistance.

Applications of the UJT

The UJT's negative resistance characteristic makes the device useful for generating repetitive signals. A circuit that is commonly used for this purpose is shown in Figure 8-22. This circuit is commonly referred to as a UJT relaxation oscillator and it is capable of generating two types of waveforms which can be used in a variety of applications.

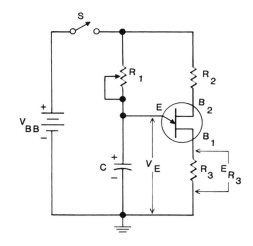

Figure 8-22 UJT relaxation oscillator.

When switch S is closed, capacitor C charges through resistor R_1 in the direction shown. When the voltage across C reaches the UJT's V_P value, the UJT turns on and the resistance between terminals E and B_1 decreases. This allows C to discharge through the UJT resistor R_3 (which has a very low value). The voltage across C quickly decreases to the UJT's V_V value and this causes the UJT to stop conducting or turn off. As soon as the UJT turns off, capacitor C starts charging again and continues until V_P is again reached. This causes the UJT to turn on which in turn allows C to discharge until V_V is reached. This action continues with the voltage across C rising slowly to V_P and decreasing rapidly to V_V. The voltage appearing between the UJT's E and B_1 terminals fluctuates as shown in Figure 8-22A. Notice that this voltage follows a sawtooth type of pattern and after the initial turn on it varies only between V_V and V_P.

Each time capacitor C discharges current is momentarily forced through R_3. These momentary pulses of current cause the voltage across $R_3(E_{R_3})$ to pulsate. These momentary voltage pulses are very narrow and a pulse occurs each time C discharges. Also note that the voltage never drops completely to zero. A small voltage will always be dropped across R_3 because a small current flows through the UJT (from B_1 to B_2) even when the device is in the off state.

The number of complete sawtooth waveforms or pulses produced each second

(frequency of oscillation) can be controlled by adjusting R_1. If the resistance of R_1 decreases, capacitor C will charge faster and the circuit will operate at a higher frequency. When R_1 increases in value, it takes longer for C to charge to V_P and the frequency decreases. The frequency can also be varied by replacing capacitor C with a larger or smaller capacitor. A larger capacitor would charge slower while a smaller capacitor would charge at a faster rate.

Figure 8-23 UJT relaxation oscillator waveforms with respect to ground at the emitter and B_1 terminals.

The sawtooth and pulse waveforms produced by the UJT relaxation oscillator can be used to perform various functions in electronic circuits. However, the UJT is also used in applications where it is not required to oscillate continuously. For example, the UJT can function as a bistable (two state) element and switch from its on state to its off state or vice-versa when it receives an appropriate input signal. The device can also be used as a frequency divider. Another important application is its use as a triggering device. The UJT is capable of producing current pulses which are ideally suited for triggering SCRs. In fact, the UJT could be used to trigger the SCR in the half-wave phase control circuit shown in Figure 8-8. In other words, the voltage across capacitor C in Figure 8-8 could be used to trigger the UJT and the UJT could in turn generate the trigger pulses needed to turn the SCR on at the proper time.

Programmable UJTs

The programmable UJT or PUT is one of the newest members of the thyristor family. This device is not constructed like an ordinary UJT (instead it uses four semiconductor layers), however it is capable of performing basically the same functions as the UJT. Also, the PUT has three terminals or leads which are referred to as the cathode, anode, and gate.

The primary difference between the ordinary UJT and the PUT is the fact that the peak voltage (V_P) of the PUT can be con-

trolled. In the UJT, the V_P value is fixed and cannot be externally controlled. The PUT's anode and cathode terminals are used in much the same way as the UJT's E and B_1 terminals and the anode is always made positive with respect to the cathode. In fact, the V-I characteristics between the PUT's anode and cathode terminals are identical to the V-I characteristics between the UJT's E and B_1 terminals. Both devices exhibit the same turn on characteristics and both have a negative resistance region between their peak and valley points. Furthermore, the V_P value of the PUT is controlled by varying the voltage between the gate and cathode terminals and the gate is always at a positive potential with respect to the cathode.

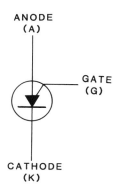

Figure 8-24 Schematic symbol for a PUT.

The operation of the PUT can be clearly demonstrated when it is connected in a circuit like the one shown in Figure 8-25. This circuit is referred to as a PUT relaxation oscillator and it produces the same basic waveforms shown in Figure 8-23. Notice that the gate-to-cathode voltage (called the gate voltage) is obtained from resistor R_4 which is part of a voltage divider network consisting of R_3 and R_4. This gate voltage is not appreciably affected by resistor R_2. R_2 has a very low value and is used to develop the output voltage pulses. As long as the gate voltage remains constant, the PUT will remain in its off or nonconducting state until its anode-to-cathode voltage (called its anode voltage) exceeds the gate voltage by an amount equal to the voltage drop across a single diode (approximately 0.7 volts). At this time the V_P value is reached and the device turns on.

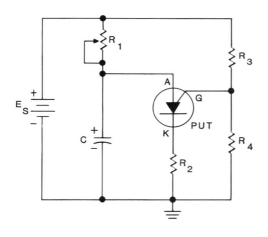

Figure 8-25 PUT relaxation oscillator.

The anode voltage is obtained from capacitor C which charges toward the supply voltage (E_S) through R_1. This is similar to the action that takes place in Figure 8-22.

Each time the voltage across C reaches the V_P value, the PUT turns on and allows C to discharge through its anode and cathode terminals and R_2. When the voltage across C drops to the V_V value of the PUT, the device turns off and the cycle is repeated. A sawtooth of voltage is developed across C and voltage pulses appear across R_2. These voltage pulses resemble the waveforms shown in Figure 8-23. As before, the frequency of oscillation can be varied by adjusting R_1, or by changing the value of C. However, in this circuit the frequency can also be varied by changing the ratio of R_3 to R_4. It is this resistance ratio that controls the PUT's gate voltage, which controls the value of V_P. If R_4 is made larger while R_3 remains constant, the gate voltage will increase and this will cause V_P to increase. The higher V_P value will make it necessary for capacitor C to charge to a higher voltage before the PUT can turn on. This will increase the time required to generate each sawtooth or pulse and decrease the frequency of operation. If R_3 increases in value while R_4 remains constant, the action will be exactly opposite and frequency will increase.

The frequency of the PUT relaxation oscillator in Figure 8-25 can be controlled by adjusting the ratio of R_3 and R_4 to control the PUT's V_P value. Although this is an extremely important consideration, it is also important to realize that resistors R_3 and R_4 controls other PUT characteristics as well. The peak current (I_P) and valley current (I_V) both depend on the values of R_3 and R_4 as well as the value of the source voltage (V_S). For example, it is possible to adjust R_3 and R_4 for any specific ratio so that any specific V_P is obtained (taking into account the value of V_S). However, it is only the ratio of R_3 and R_4 that determines V_P, not the individual values of the resistors. The actual resistor values determine the PUT's I_P and I_V values for any given value of V_S.

This relationship between the external voltage divider network (R_3 and R_4) and the values of I_P and I_V is somewhat complex. However, the relationship can be simplified by relating I_P and I_V to the combined values of R_3 and R_4. R_3 and R_4 are considered to present an equivalent resistance (R_G) to the gate of the PUT which is equal to their parallel equivalent:

$$R_G = \frac{R_3 \, R_4}{R_3 + R_4}$$

For any given value of V_S, I_P and I_V both will decrease as R_G increases. More complex equations may also be used to determine the exact values of I_P and I_V under various circuit conditions or these values can be determined graphically when appropriate charts or curves are provided. The circuit designer may use either of these methods to calculate the required values and in general all of the quantities just described must be considered when designing an oscillator circuit like the one shown in Figure 8-25. The intent here is not to show a complete design procedure for a PUT oscillator; but to emphasize the PUT's important electrical characteristics.

In addition to its use in relaxation oscillators, the PUT may also be used as a triggering device for SCRs. Due to its four-layer construction the PUT is capable of supplying trigger pulses that have a higher amplitude than those obtained from ordinary UJTs and can therefore trigger SCRs that have higher current ratings. The PUT is also used in long duration timer circuits (which are basically identical to the circuit in Figure 8-25) because the device has a low gate-to-anode leakage current. This is important in timer circuits since this leakage current adds to the charging current flowing through the capacitor. In such a circuit the capacitor must charge over a long period of time before the V_P value is reached and an additional charging current (leakage) cannot be tolerated.

The PUTs important electrical characteristics can be controlled, thus making it more versatile than an ordinary UJT. The PUT is also more sensitive than a UJT and it responds faster. It is therefore used in place of the UJT in a number of applications although the UJT is still preferred in certain types of circuits.

27. The UJT has one semiconductor junction.

    A. True
    B. False

28. The leads attached to each end of the UJT's N-type bar are referred to as _____ and _____.

29. The lead that is attached to the UJT's P-type pellet is referred to as the _____.

30. A PUT has _____ PN junctions.

31. An equivalent UJT circuit can be formed with two resistors (R_{B_1} and R_{B_2}) and a _____.

32. The ratio of R_{B_1} to the total resistance ($R_{B_1} + R_{B_2}$) is called the _____ _____.

33. The input voltage value (V_E) required to turn on a UJT is referred to as the _____ _____ and is designated as V_p.

34. The current that flows through the UJT's emitter lead at the point where V_p is reached is referred to as the _____ _____ and is designated as I_p.

35. Beyond the V_p point V_E decreases until it reaches a minimum value which is referred to as the _____ _____ and is designated as V_V.

36. The current that flows through the UJT's emitter lead at the V_V point is referred to as the _____ _____ and is designated as I_V.

37. Between V_p and V_V, the UJT exhibits a _____ resistance.

38. When used as a relaxation oscillator, the UJT can generate both sawtooth and pulse type waveforms.

    A. True
    B. False

39. The UJT is often used as a triggering device for the _____.

40. The PUT can perform the same basic functions as a UJT although it is constructed differently and has three leads which are referred to as the _____, _____ and _____.

41. The PUT's V_p value can be controlled by changing the voltage between its _____ and _____.

42. In Figure 8-25 the PUT's Vp value is varied by changing the ratio of _____ and _____.

43. If R_3 increases in value while R_4 remains constant, the circuit in Figure 8-25 will operate at a _____ frequency.

Unit Summary

The SCR is the most popular member of the thyristor family. This device has four alternately doped semiconductor layers and three leads which are referred to as the anode, cathode, and gate. This device acts like an electronic switch which can be turned on by a momentary gate voltage and turned off by reducing its operating current below a certain holding value. The SCR may be used to switch DC or AC power and it can even be used to vary the amount of AC power applied to a load. The SCR conducts current in one direction only.

The triac effectively performs the same function as two SCRs that are connected in parallel but in opposite directions. It can therefore control AC currents more efficiently than a single SCR since it operates on both the positive and negative portions of each AC input cycle. The triac must be used in conjunction with a triggering device because of its nonsymmetrical triggering characteristics. The triac conducts current in both directions, but it is not capable of the higher currents that the SCRs can handle.

The triggering device most commonly used with the triac is known as a diac. The diac is simply a bidirectional solid-state switch that turns on in either direction, when its breakdown voltage in each direction is exceeded. The diac has symmetrical switching characteristics since its breakdown voltage is approximately the same in each direction.

When the diac is used to trigger a triac, it forces the triac to turn on each time it turns on. This effectively compensates for the nonsymmetrical characteristics of the triac.

Although it has characteristics that are different from those of SCRs, triacs, and diacs, the unijunction transistor (UJT) is still considered to be a thyristor. This device has only one PN junction and three leads (base 1, base 2, and the emitter). The UJT's important electrical characteristics occur between its emitter and base 1 leads. The device will turn on and conduct a substantial current through these leads when the voltage across these leads reaches a certain maximum value known as the peak voltage (V_p). Beyond this point the device exhibits a negative resistance until a certain minimum voltage known as the valley voltage (V_V) is reached. These characteristics allow the UJT to generate repetitive signals when used in a relaxation oscillator circuit or it can be used as a bistable switching element or as a triggering device.

A special type of UJT known as a programmable UJT or PUT is also available. This device operates like an ordinary UJT but its Vp value as well as other important electrical characteristics can be varied by an external control voltage. This makes the PUT more versatile than the UJT and therefore more suitable for a broader range of applications. The PUT is widely used as a triggering device and it is often used in long duration timer circuits.

Unit 9
Integrated Circuits

Contents

Introduction

In this unit you will examine a relatively new type of solid-state device that is known as an integrated circuit or IC. The integrated circuit is actually a group of extremely small solid-state components which have been formed within or on a piece of semiconductor material and then interconnected to form a complete circuit. The IC is therefore a solid-state circuit and not an individual solid-state component like a diode or a transistor.

Integrated circuits are constructed in basically four different ways. They can be produced as monolithic, thin-film, thick-film, or hybrid devices. IC's may also be divided into two major groups according to their mode of operation. They are designated for either linear operation or digital operation.

The integrated circuit is an extremely important solid-state device which is now widely used throughout the electronics industry. You must be familiar with this important device if you plan to service, design, or intimately work with almost any type of electronic equipment. This unit explains why integrated circuits are used, how they are constructed, and where they find their greatest applications. Study this unit carefully, as it contains information that will help you throughout your electronics career.

When you have completed this unit on integrated circuits you should be able to:

1. State the need for integrated circuits in electronics.

2. Name at least three advantages that integrated circuits have over conventional circuits.

3. Name three disadvantages of using integrated circuits as compared to conventional circuits.

4. Explain the difference between monolithic, film-type, and hybrid integrated circuits.

5. Explain the difference between linear and digital integrated circuits.

6. Recognize three basic integrated circuit packages.

7. Explain the difference between SSI, MSI, and LSI circuits.

The Importance Of Integrated Circuits

Integrated circuits, or IC's have changed the entire electronics industry. Before IC's were developed, all electronic circuits consisted of individual, or discrete, components that were wired together, often requiring a large amount of physical space. Printed circuit board technology made it possible to reduce the amount of space required. However, electronic circuits can be quite complex, requiring a large number of components, and since discrete components have a fixed physical size, there is a practical limitation on the amount of size reduction that can be achieved. Therefore, an important consideration in the design of circuits containing discrete components is the amount of space required by the circuit. However, the development of integrated circuit technology has made it possible to fabricate large numbers of electronic components on a single silicon chip. As a result, the physical size of a circuit can be significantly reduced, making it possible to design circuits and devices that would otherwise be impractical.

IC's are complete electronic circuits consisting of as many transistors, diodes, resistors, and capacitors as may be necessary for circuit operation. They are encapsulated in packages that are often no larger than a single transistor. The technology and materials used in the manufacture of IC's are basically the same as those used in the manufacture of transistors and other solid-state devices. In addition, IC's are manufactured for a wide variety of applications and, as a result, are used throughout the electronics industry. This trend will probably continue as manufacturers and engineers find more and more applications for this versatile family of devices.

Advantages

The small size of the integrated circuit is its most apparent advantage. A typical IC can be constructed on a piece of semiconductor material that is less than one tenth of an inch square. Even when the IC is suitably packaged, it still occupies only a small amount of space. The first standard IC package was approximately one quarter of an inch long and one eighth of an inch wide. However, it later became apparent that in most applications this package was smaller than was actually necessary because the final size of the equipment was often dictated by other components which were much larger. Therefore, the IC's that are used today come in packages that are somewhat larger than the first ones that were developed.

Integrated circuits have been used extensively in the aerospace industry to reduce the size and weight of satellites, missiles, and other types of space vehicles. They have helped to reduce the size of complex computer systems and they are used in devices such as hearing aids and portable electronic calculators. In most applications where size and weight must be reduced to an absolute minimum, the IC can make a significant contribution.

The small size of the integrated circuit also produces other fringe benefits. The smaller circuits consume less power than conventional circuits and they cost less to operate. They generate less heat and therefore generally do not require elaborate cooling or ventilation systems. The smaller circuits are also capable of operating at higher speeds because it takes less time for signals to travel through them. This is an important consideration in the digital field where thousands of decision-making circuits are used to provide rapid solutions to a variety of problems.

The integrated circuit is also more reliable than a conventional circuit that is formed with discrete components. This greater reliability results because every component within the IC is a solid-state device and these components are permanently connected with thin layers of metal. They are not soldered together like the components in a conventional circuit and a circuit failure due to faulty connections is less likely to occur. Also, all of the components within the IC are simultaneously formed as opposed to conventional circuitry which is assembled in a step-by-step manner. Therefore, there is less chance of making mistakes during the assembly of an IC.

Integrated circuits are also thoroughly tested after they are assembled and only those devices which meet the required specifications are considered suitable for either military and space or industrial and commercial applications. This extensive testing of IC's combined with the construc-

tion techniques previously described, produce a device that is highly reliable. In fact, when the IC is compared with conventional circuitry, it is often found to be thirty or even fifty times more reliable. This high degree of reliability has made the IC an important component for use in aerospace equipment and in complex digital computer systems. In each of these applications, a tremendous amount of circuitry is used and this circuitry must be reliable. Aerospace equipment, such as satellites and missiles, are sometimes required to operate for years without failure and a highly complex digital computer can be rendered inoperative if just one component fails to operate properly.

The use of integrated circuits can also result in a substantial cost savings. When IC's are manufactured in large quantities, they can often be sold at prices which are well below those of conventional circuits. Manufacturers of IC's usually offer a standard line of devices which are produced in large quantities. If designers can utilize standard IC's in their equipment and if they purchase them in large quantities, they can often save a considerable amount of money. However, there are situations where the cost of an integrated circuit can be higher than that of a conventional circuit. This can occur when the designer needs a special purpose IC or one that is specially constructed by the IC manufacturer to suit the needs of the designer. Under these conditions, the cost might be prohibitive, especially when only a few IC's are purchased. However, as the IC manufacturers gain experience and manufacturing techniques improve, the cost of most types of IC's continue to decrease and this trend is likely to continue.

By using integrated circuits in electronic equipment, additional cost savings can also be realized. Any equipment which utilizes IC's, instead of conventional circuitry, will have a fewer number of parts which must be assembled. Therefore, less wiring is required and less time is needed to assemble the equipment. This can mean a considerable cost reduction when a large number of units are to be produced. Also, the equipment manufacturer only has to procure and stock a relatively small number of IC's as compared to the relatively large inventory of discrete components that would be required if conventional circuitry was used. All of these factors can result in a reduction in overhead and cost savings for the equipment manufacturer, which can be passed on to the consumer.

Disadvantages

It might appear that the integrated circuit has only advantages to offer and no real disadvantages. Unfortunately, this is not the case, since IC's do indeed have certain limitations which make them unsuitable for certain applications. Since the IC is an extremely small device, it cannot handle large currents or voltages. High currents generate heat within the device and the tiny components can be easily damaged if the heat becomes excessive. High voltages can break down the insulation between the components in the IC because the components are very close together. This can result in shorts between adjacent components which would make the IC completely useless. Therefore, most IC's are low power devices, which have low operating currents (in the milliampere range) and low operating voltages (5 to 20 volts). Also, most IC's have power dissipation ratings of less than 1 watt.

At the present time, only four types of components are commonly constructed within an IC, thus making available only a narrow selection of components. Diodes and transistors are the easiest to construct and are used extensively to perform as many functions as possible within each IC. Resistors and capacitors may also be formed, but they are much more difficult to construct. The amount of space occupied by a resistor increases with its value and in order to conserve space, it is necessary to use resistors with values that are as low as possible. It is also difficult to control the exact values of resistors, although the ratio between resistor values can be controlled with a high degree of accuracy. Capacitors occupy even more space than resistors and the amount of space required increases with the value of the capacitor. Therefore, capacitor values are made as small as possible and capacitors are used only when it is absolutely necessary. For these reasons, you will see diodes and transistors used extensively in IC's while resistors and capacitors are used sparingly.

Integrated circuits cannot be repaired because their internal components cannot be separated. When one internal component becomes defective, the whole IC becomes defective and must be replaced. This is definitely a disadvantage because it means that good components must be thrown away with the bad and it also means that there is the additional expense of replacing an entire circuit. However, this disadvantage is not as bad as it might first appear because it is offset by other factors. First of all, the task of locating a trouble within a system is simplified because it is only necessary to trace the problem to a specific circuit instead of an individual component. This greatly simplifies the task of maintaining highly complex systems and reduces the demands that are placed on the maintenance personnel. Also, it is possible to reduce the time required to repair the equipment and the spare parts inventory can usually be smaller.

When all of the factors are considered, the disadvantages associated with the use of integrated circuits are outweighed by their advantages. There is a definite need for integrated circuits in the electronics industry. IC's are making it possible to reduce the size, weight, and cost of electronic equipment but at the same time increase its reliability. As manufacturing techniques improve, IC's are becoming more sophisticated and are capable of performing a wider range of functions. Therefore, in the future, these devices will undoubtedly find an increasingly broader range of military, space, industrial, and commercial applications.

1. An integrated circuit can be constructed on a piece of semiconductor material that is less than one tenth of an inch square.

   A. True
   B. False

2. Since the IC is extremely small it consumes only a small amount of _____.

3. When IC's are used, the _____ and _____ of electronic equipment can be reduced.

4. Integrated circuits are capable of operating at higher _____ than equivalent circuits using discrete components.

5. Since the IC is a single solid-state device, it is more _____ than a conventional circuit which uses a variety of discrete components.

6. When IC's are manufactured in large quantities they usually cost _____ than equivalent circuits using discrete components.

7. Integrated circuits cannot handle high operating:

   _____ or _____.

8. The two IC components that are the easiest to construct are _____ and _____.

9. The two IC components that are the most difficult to construct are _____ and _____.

10. When a single component within an IC becomes defective, the entire IC must be replaced.

    A. True
    B. False

As mentioned earlier, there are basically four types of integrated circuits. They are classified as monolithic, thin-film, thick-film, and hybrid devices. Each type has certain advantages over the others and also certain limitations. We will now see how each of these basic IC's are fabricated and we will also examine the most popular IC packages that are used.

Monolithic IC's

The monolithic IC is constructed in basically the same manner as a bipolar transistor, although the overall process requires a few additional steps because of the greater complexity of the IC. The fabrication of monolithic devices begins with a circular semiconductor wafer (usually silicon) as shown in Figure 9-1A. This wafer is usually very thin (.006 to .012 inch thick) and may be as small as one inch or as large as two inches in diameter. The semiconductor wafer serves as a base on which the tiny integrated circuits are formed and is commonly referred to as a substrate.

Many IC's are simultaneously formed on a single wafer as shown in Figure 9-1B. Each square shown in this figure represents a single IC. The number of IC's formed on a wafer will depend on the size of the wafer and the size of the individual IC's. In some cases a wafer will contain less than 300 IC's but in other cases it may contain 500 to 600 IC's or possibly even more. All of the IC's formed on a wafer are usually the same type

and therefore have the same physical dimensions. Also, each IC contains the same number and type of components.

When all of the IC's have been simultaneously formed, the wafer is sliced into many sections as shown in Figure 9-1C. These sections are commonly referred to as chips or dice. Each chip represents one complete integrated circuit and contains all of the components and wiring associated with that circuit. Once the IC's are separated into individual chips, each IC must be mounted in a suitable package and tested. Throughout this entire fabrication process some IC's are rejected because they have various physical and electrical deficiencies. Although an extremely large number of IC's are fabricated in any given process, only a relatively small number of these devices are found to be usable. The number of usable devices is usually expressed as a percentage of the maximum number that it is possible to obtain and is referred to as the yield. The yield in a typical fabrication process may be less than 20 percent. This simply means that less than 20 percent of the IC's produced in a particular process are found to be acceptable.

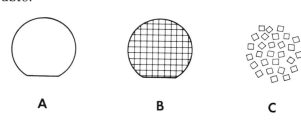

Figure 9-1 Integrated circuit construction.

Over 200 monolithic IC's have been formed on this semiconductor wafer. The wafer is ready to be sliced into many individual IC's which will be subsequently packaged and tested.

Now that we have briefly reviewed the over-all IC construction process, let's take a closer look at an individual monolithic IC and see exactly how the individual components within the IC are formed. As we examine the various IC components, you must remember that these components are formed within a number of IC's simultaneously on a semiconductor wafer. A single monolithic IC is never constructed alone.

Bipolar IC's

As mentioned previously, the components that are commonly used in IC's are diodes, transistors, resistors, and capacitors. These components can be formed by diffusing impurities into selected regions of a semiconductor wafer (substrate) to produce PN junctions at specific locations. The basic manner in which these four components are formed and the manner in which they are interconnected are shown in Figure 9-2.

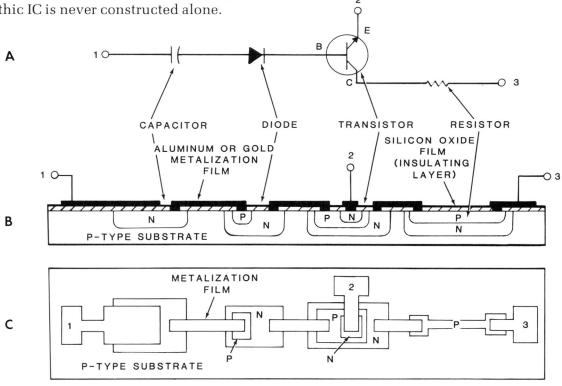

Figure 9-2 Basic construction of a single integrated circuit.

Definition

Diffusion: A popular construction technique which allows a gaseous N- or P-type impurity element, under high temperature, to penetrate a semiconductor wafer to produce a respective N- or P-type region.

Figure 9-2A shows a simple electronic circuit consisting of a capacitor, a PN junction diode, an NPN transistor and a resistor. Operating voltages and currents can be applied to the circuit through terminals 1, 2, and 3 as shown. This circuit could be easily constructed using four discrete components, however, it can also be produced as a monolithic integrated circuit. When produced in IC form, the circuit will appear as shown in Figure 9-2B and Figure 9-2C.

Observe the IC structure shown in Figure 9-2B and compare it with the circuit in Figure 9-2A. Notice that all four of the components are formed within a P-type substrate or wafer. These components are simultaneously formed by diffusing N-type and P-type impurities into the P-type substrate to produce N-type or P-type regions. Several diffusion operations are necessary to form the entire IC and it is necessary to accurately control the location, the size, and the depth of each N-type or P-type region that is formed. This is accomplished by depositing an insulating layer of silicon oxide on top of the substrate. Then appropriate windows are cut in the oxide coating by means of an acid so that only the desired areas on the substrate are exposed. Then N-type impurities are diffused through the windows and into the substrate to form the first and largest N-type regions. Next, the windows are covered with new oxide and new windows are formed. Then a P-type impurity is diffused into the substrate at the appropriate locations to form P-type regions. This process is repeated again to form the final N-type regions.

Notice that the NPN transistor in Figure 9-2B is constructed in basically the same manner as a conventional transistor. Its N-type collector was formed first. Then its P-type base and N-type emitter regions were formed. All three regions extend to the top of the substrate and lie in a flat plane, thus giving the device a planar type of structure. The transistor in Figure 9-2B is therefore equivalent to a conventional planar transistor that is formed with the diffusion method.

The PN junction diode in Figure 9-2 is also equivalent to a conventional diode. The larger N-type region, that was first diffused into the substrate, serves as the cathode and the smaller P-type region, that was diffused into the N-type region, serves as the anode.

The resistor is formed by first producing a large N-type region and then forming a P-type region on top of it. This produces a long narrow strip of P-type material that is surrounded by N-type material. This long P-type region serves as a resistor and its resistance can be controlled by adjusting its length and width. Increasing the length will result in a higher resistance, while increasing the width will lower the resistance. The resistance value can also be adjusted by controlling the concentration of impurities within the P-type strip. A higher concentration of impurities will result in a lower resistance and vice-versa.

In general, the amount of space required to produce a resistor increases with its value, thus making it difficult to place large values within an IC. Also, it is difficult to produce resistors with highly accurate values. In general it is difficult to obtain resistance tolerances that are much better than plus or minus 10 percent, thus making it impossible to construct certain types or circuits in IC form. However, the ratio between two resistors that are formed close together on a substrate can be regulated with a high degree of accuracy. In fact, resistance ratios that are accurate to within plus or minus 1 percent can be obtained. Therefore, circuits that are produced in IC form are generally designed to take advantage of the accurate resistance ratios that are available, instead of specific resistance values.

The capacitor in Figure 9-2B is constructed in a unique manner. The diffused N-type region serves as the bottom plate of the capacitor and the layer of silicon oxide serves as the dielectric. The top plate of the capacitor is simply a layer of metal that has been deposited on top of the oxide coating. This type of capacitor is sometimes referred to as a metal-oxide capacitor.

The value of capacitance is determined by the area of the plates, the thickness of the oxide layer, and the dielectric constant of the oxide layer. In general, the amount of capacitance obtainable per unit area is very low and it is necessary to make the capacitor quite large to obtain a significant value of capacitance. In most cases, it is not practical to construct capacitors with values that are much higher than a few hundred picofarads. Even the smaller 10 and 20 picofarad capacitors require substantially more space than a transistor.

Another type of capacitor is also used in monolithic IC's. This device makes use of the capacitance that exists across the depletion region within a reverse-biased PN junction and it is commonly referred to as a junction capacitor. The junction capacitor is constructed in basically the same manner as the PN junction diode shown in Figure 9-2B. The amount of capacitance obtainable within the junction capacitor per unit area is also very low and, in general, the maximum capacitance values that can be obtained are somewhat lower than with metal-oxide devices. Also, the capacitance changes when the reverse bias voltage changes. This is because the bias voltage controls the size of the depletion region. This means that it is always necessary to maintain the proper bias voltage across the device.

As previously mentioned, all of the components in Figure 9-2B are simultaneously formed by coating the substrate with silicon oxide, cutting appropriate windows into the oxide coating, and diffusing impurities into the substrate. This process is repeated several times until all components are formed.

In the final steps of the process, an oxide film is deposited over the substrate and windows are cut into the oxide to expose the various regions within each component. Then a layer of aluminum or gold is deposited over the oxide and allowed to penetrate and make contact with the exposed regions. Next, acid is used to etch away certain portions of the metal film, leaving behind narrow strips of metal which serve as conductors and interconnect the various components to form a complete circuit. The end result of this operation can be seen in Figure 9-2B. A top view of the IC (with the silicon oxide removed) is also shown in Figure 9-2C. Notice that the four components are connected together by thin metal strips. The three terminals which provide access to the integrated circuit are also shown. Terminal 1 is simply the top plate of the capacitor which is formed simultaneously with the other metal strips. Terminals 2 and 3 are just metal strips which make direct contact with the transistor and resistor, respectively. As shown in Figure 9-2C, the metal strips which connect to terminals 1, 2, and 3 are brought out to the perimeter of the IC and enlarged to form rectangular pads. These pads (identified as 1, 2, and 3) provide relatively large metal surfaces to which fine metal wires can be bonded when the IC is permanently installed in a suitable package.

The components in the monolithic IC just described are formed by using the same basic construction techniques used to produce bipolar transistors. Integrated circuits that are produced in this manner are often called bipolar integrated circuits. Bipolar IC's are the most popular type of IC and are used in a wide variety of applications.

The monolithic IC in Figure 9-2 is extremely simple when compared to most other units that are constructed. Most IC's contain a large number of components and are therefore quite complex. In general, IC's are classified according to their complexity or the number of components that they contain. The terms small-scale integration (SSI), medium-scale integration (MSI), and large-scale integration (LSI) are commonly used to identify these circuits. Although these terms have never been exactly defined, SSI devices usually have less than 200 components while MSI devices generally have between 200 and 1000 components. When the IC has over 1000 components it is usually referred to as an LSI device. The simple IC shown in Figure 9-2 is an SSI device.

As you examine Figure 9-2, you may wonder why the various components do not interface with each other when current flows through the circuit. Actually, the components are electrically isolated or insulated from each other due to the unidirectional characteristics of the PN junctions that are used. Due to the diffusion techniques that are used, each component is formed within its own N-type region and therefore each component is separated from the P-type substrate by a PN junction. When operating, the circuit is biased so that the P-type substrate is more negative than any other part of the circuit. This causes each PN junction to be reverse-biased and offers a high resistance which isolates the various components. The operating currents that flow through the circuit are forced to take the proper paths. When IC's are constructed in this manner, they effectively utilize diode isolation as a means of electrically separating the various components.

Another construction technique is also sometimes used which make use of oxide-isolation. With this method, a layer of silicon oxide is actually formed around each component, thus providing an insulating layer between each component and the substrate. A third method, known as beam-lead isolation, is also used in special applications. With this method, the components are formed as described earlier in the text, but a heavier metalization is used to form the interconnections and then the semiconductor material between the components is removed. The components are completely separated but supported by the heavy metalization that serves as the interconnecting conductors. A special type of plastic is then used to fill the spaces that were previously occupied by semiconductor material. With this method, an extremely high degree of isolation can be obtained.

MOS IC's

To this point we have assumed that the transistors used in IC's are always bipolar devices. However, this is not the case. The fact is, IC's are often designed to utilize either bipolar transistors or field-effect transistors. The FET's that are used in IC's are the insulated-gate devices which are known as IGFET's or MOSFET's. These MOSFET's can be either N-channel or P-channel devices and they may operate in either the depletion or the enhancement mode.

The MOSFET's used in IC's are basically constructed like the conventional MOSFET's described in an earlier unit. A typical integrated circuit MOSFET is shown in Figure 9-3. Notice that the source and drain regions are diffused into the substrate. These regions will always be doped oppositely with respect to the substrate. A thin layer of silicon oxide is formed over the substrate and appropriate windows are cut into it so that the metal electrodes (terminals) can be formed at the proper locations.

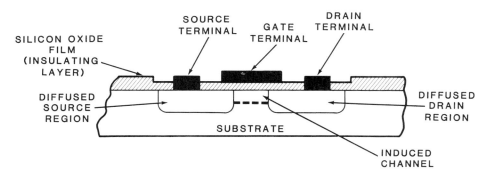

Figure 9-3 A typical integrated circuit MOSFET.

Notice that the gate terminal is separated from the substrate by an extremely thin oxide layer. This oxide layer may be only 1000 angstroms thick (1 angstrom equals 1×10^{-10} meters), but it completely isolates the gate from the substrate. When properly biased, current will flow from source to drain through a channel that is formed directly under the gate, thus causing the device to operate in the enhancement mode.

The source, drain, and gate regions within the MOSFET can be formed very close together. In fact, a typical MOSFET can be made much smaller than a bipolar transistor. This fact alone makes the MOSFET a very desirable component for use in integrated circuits. However, these devices have another distinct advantage. Due to the manner in which they are constructed, a high degree of isolation between components can be obtained. The gate is completely isolated by the oxide coating and the source and drain regions are isolated by PN junctions.

The MOSFET can also be used as a resistor when it is properly biased. The resistance of the channel that is formed between the source and drain terminals can be adjusted by regulating its operating voltages. This resistance is also determined by the transconductance of the device. When MOSFET's are used as resistors, a wide spacing is used between the source and drain regions to obtain a lower transconductance and the gate is connected directly to the drain so that the device is always conducting.

Integrated circuits which use MOSFET's as the principle controlling elements are generally referred to as MOS IC's. Many MOS IC's contain only MOSFET's while others contain combinations of MOSFET's and MOSFET resistors. MOS IC's are produced with the same basic technology that is used to produce field effect transistors, while bipolar IC's are based on bipolar transistor technology.

Since MOS components are smaller than bipolar components, the MOS circuit can be constructed in a smaller space than an equivalent bipolar circuit. This higher component density in MOS circuits make them highly suitable for use in MSI and LSI circuits. MOS circuits also consume less power, provide greater temperature stability, and offer a higher input impedance than comparable bipolar circuits. Unfortunately, they have one important disadvantage which limits their use. In general, MOS circuits cannot respond as quickly as bipolar circuits. When used in digital or logic applications or in any other applications where basic switching actions are required, they cannot operate as fast as bipolar circuits. Even when used to amplify or generate electronic signals, MOS circuits cannot operate at the high frequencies which are obtainable with bipolar circuits. Therefore, in certain applications bipolar IC's may be more desirable than MOS circuits while in other cases the exact opposite will be true. With the advancements that have been made in MOS technology recently, the speed problem may be eliminated in the near future.

Thin-Film IC's

Unlike monolithic IC's, which are formed within a semiconductor material (substrate), the thin-film circuit is formed on the surface of an insulating substrate. In the thin-film circuit, components such as resistors and capacitors are formed from extremely thin layers of metals and oxides which are deposited on a glass or ceramic substrate. Interconnecting wires are also deposited on the substrate as thin strips of metal. Components such as diodes and transistors are formed as separate semiconductor devices and then permanently attached to the substrate at the appropriate locations.

The substrate on which the thin-film circuit is formed is usually less than one inch square. The resistors are formed by depositing tantalum or nichrome as the thin films or strips on the surface of the substrate. These films are usually less than 0.0001 inch thick. The value of each resistor is determined by the length, width and thickness of each strip that is formed on the substrate. Several thin-film resistors and their interconnecting conductors are shown in Figure 9-4. The short, wide resistor has a relatively low value while the longer and narrower resistors have higher values. The interconnecting conductors are extremely thin metal strips which have been deposited on the substrate. Low resistance metals, such as gold, platinum, or aluminum, are generally used as conductors. The substrate is made from an insulating material that will provide a rigid (nonflexible) support for the components. Glass or ceramic materials are often used as substrates.

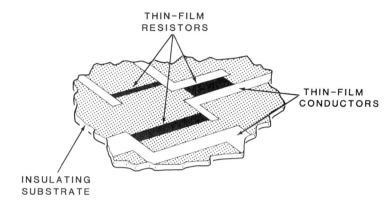

Figure 9-4 A portion of a thin-film circuit showing resistors and conductors.

Using the thin-film techniques, it is possible to produce extremely accurate resistance values. The tolerances on resistor values can be as low as ± 0.1 percent over a resistance range that extends from several ohms to more than 100 kohms. Furthermore, it is possible to obtain extremely accurate ratios between various resistors in a thin-film circuit. Ratio accuracies of ± 0.01 percent can be readily obtained.

Thin-film capacitors consist of two thin layers of metal separated by an extremely thin dielectric layer. One metal layer is deposited on the substrate and then an oxide coating is formed over the metal to serve as a dielectric. Next, the top plate of the capacitor is formed by depositing a thin metal film over the dielectric. Materials such as tantalum, gold or platinum can be used to form the plates and the dielectric may be formed from insulative materials such as tantalum oxide, silicon oxide, or aluminum oxide. The required capacitance value is obtained by adjusting the area of the plates and by varying the thickness as well as the type of dielectric material used.

When diodes or transistors are required in a thin-film circuit, they are produced as separate semiconductor components using the same basic monolithic techniques that are used to form conventional transistors. Many of the diodes and transistors that are used in thin-film circuits are formed by diffusion methods and have a planar type of construction. The diode and transistor chips are permanently mounted on the thin-film substrate and then electrically connected to the thin-film circuit with extremely thin wires.

The materials used in the construction of thin-film components and conductors are usually deposited on an insulating substrate by using either an evaporation or sputtering process. In the evaporation process, the materials are placed within a vacuum and heated until they evaporate. The resulting vapor is allowed to condense on the substrate to form a thin film. The sputtering process takes place in a gas filled chamber. High voltages are used to ionize the gas and the material is bombarded with ions. The atoms within the material become dislodged and drift toward the substrate where they are deposited in the form of a thin film. To insure that the proper films are deposited at the proper locations on the substrate, various masks can be used which will expose only the desired regions, or the substrate

can be completely coated with a film and then the undesired portions can be cut or etched away.

Thick-Film IC's

Thick-film IC's are formed in a somewhat different manner than the thin-film devices just described. In thick-film IC's, the resistors, capacitors and conductors are formed on an insulating substrate by using a silk-screen process. In the silk-screen process, a very fine wire screen is placed over the substrate and a metalized-ink is forced through the screen with a squeegee. Only certain portions of the wire screen are open, thus allowing the ink to penetrate and coat specific portions of the substrate. The remaining holes in the screen are filled with a special emulsion. In this manner, a pattern of interconnecting conductors is formed on the substrate. The pattern is then heated to over 600 degrees centigrade and the painted surfaces harden and become low resistance conductors. Resistors and capacitors are also silk-screened on top of the substrate by forcing the appropriate materials (in paste form) through an appropriate screen and then heating the substrate to a high temperature. This process is repeated using various pastes until the circuit is complete except for diodes or transistors. As with thin-film circuits, these components must be formed separately as semiconductor devices and then added to the substrate to complete the circuit.

Subsequent operations may also be required

after the silk-screening is completed. For example, it is generally necessary to trim the resistors to obtain accurate resistance values. This trimming operation can be performed with an air-abrasive (sand blasting) technique or a laser can be used to burn away excessive resistor material. This trimming operation can produce resistors with tolerances as low as ± 0.5 percent and these resistors can be produced in standard values that can extend from 5 ohms to 100 megohms. Thick-film capacitors have relatively low values which can typically range up to several thousand picofarads. When higher capacitance values are required, the film type capacitors cannot be used. To obtain higher capacitance values, it is necessary to use miniaturized discrete capacitors and mount them permanently on the substrate.

The thick films formed in the silk-screening process are usually more than 0.0001 inch thick. Thick-film components are therefore larger than the thin-film components described earlier and come closer to resembling conventional discrete components. The tolerances involved in the construction of thick-film circuits are therefore not as critical as those associated with thin-film circuits. Thin-film components actually have no discernible thickness and in many cases appear to be simply painted or printed on a substrate.

Hybrid IC's

Hybrid integrated circuits are formed by utilizing various combinations of monolithic, thin-film and thick-film techniques and they may even contain discrete semiconductor components in chip form. Therefore, many types of hybrid circuit arrangements can be produced. A typical hybrid circuit might consist of a thin-film circuit on which various monolithic IC's have been attached or it could utilize monolithic IC's thick-film components and discrete diodes and transistors that are all mounted on a single insulating substrate.

A portion of a typical hybrid IC is illustrated in Figure 9-5. An insulating substrate is used to support the circuit components as shown. Notice that a monolithic IC is mounted on the substrate along with thick-film resistors and a small discrete capacitor. All of the components are interconnected with conductors that are formed on the substrate using film techniques. The monolithic IC is connected to the conductors with fine wires that are bonded in place. The thick-film resistors will usually have notches cut into them, as shown, as a result of the trimming process that is used to adjust their values. The capacitors used in these circuits can be formed either by using film techniques or miniature devices can be installed between conductors as shown. The conductors are routed to the outer edge of the substrate, where they are enlarged to form circular access terminals for the circuit. When the hybrid circuit is packaged, suitable pins are soldered to these terminals so that the circuit can be plugged into a matching receptacle or socket.

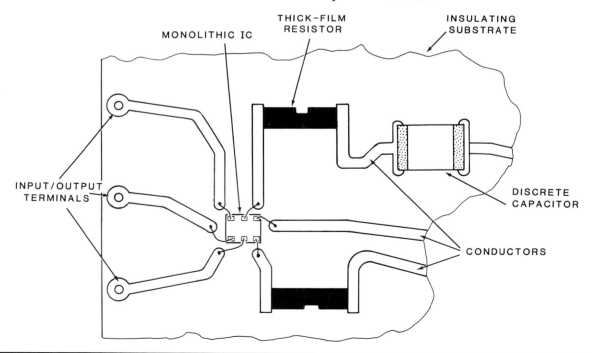

Figure 9-5 A portion of a typical hybrid IC.

The hybrid approach to the construction of IC's allows a high degree of circuit complexity (a large number of components) by using monolithic circuits and at the same time takes advantage of the extremely accurate component values and tolerances that can be obtained with film techniques. Discrete components such as diodes, transistors, or capacitors are often used in hybrid circuits because they can handle relatively large amounts of power and because they exhibit certain desirable electrical characteristics that are difficult to achieve in a single monolithic circuit.

If only a few circuits are required, it can be much cheaper to utilize hybrid IC's instead of monolithic IC's. This is because the construction of hybrid circuits does not involve the elaborate and expensive circuit layout design procedures and diffusion techniques associated with monolithic circuits. The major expense in hybrid circuit construction is in the wiring and assembly of the components and in final packaging of the devices. Therefore, hybrid circuits can be more easily designed and packaged for special applications. However, when a large number of circuits are required, monolithic devices are usually the better choice, provided that they can perform the required functions. Monolithic circuits are considerably less expensive when they are manufactured and sold in large quantities.

Since hybrid circuits utilize discrete components as well as monolithic and film circuits, they are often larger and heavier than monolithic IC's. The use of discrete components also tends to make them somewhat less reliable than monolithic circuits.

IC Packages

Like transistors and other types of solid-state components, IC's are mounted in packages which protect them from moisture, dust and other types of contaminants. The packages also make it easier to install the IC's in various types of equipment, since each package contains leads which can be either plugged into matching sockets or soldered to adjacent components or conductors.

Many different types of IC packages are available and each type has its own advantages and disadvantages. The most popular IC package is the dual in-line package, which is commonly referred to as a DIP. A typical dual in-line package is shown in Figure 9-6. Notice that the package has two rows of mounting pins or leads which can be inserted in a matching socket, or inserted in holes on a printed circuit board and soldered in place. The package shown has 12 leads in each row or a total of 24 pins. However, both larger and smaller dual in-line packages are available to accommodate the wide range of SSI, MSI, and LSI circuitry.

Figure 9-6 A typical dual in-line integrated circuit package.
(Courtesy of Mostek)

Dual in-line packages may be constructed from either plastic or ceramic materials. Plastic devices are relatively inexpensive and are considered suitable for most commercial and industrial applications where operating temperatures fall within a range of 0 to 70 degrees centigrade. The ceramic devices are somewhat more expensive but offer better protection against moisture and contaminants. They can also withstand a wider range of operating temperatures (-55 to $+125$ degrees centigrade) and are generally recommended for use in military, aerospace, and severe industrial applications. Some of the most popular DIP's are made of plastic and have a total of 14 or 16 leads and some smaller versions with only 8 leads are also used. These smaller devices are commonly referred to as miniature dual in-line packages or mini DIP's. Several mini DIP's are shown in Figure 9-7.

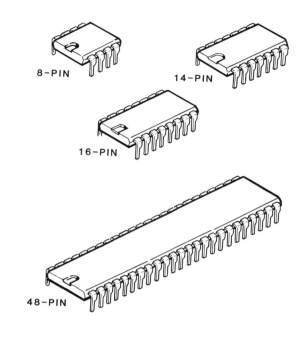

Figure 9-7 Typical miniature dual in-line packages.

Dual in-line packages are most commonly used with monolithic integrated circuits, but they can also be used with other types of IC's. The tiny IC chips are permanently mounted inside their respective packages and electrical connections between the chips and package leads are made with fine wires that are bonded in place. Although it is common practice to use one IC chip in a package, some complex circuits must be formed by mounting several chips in a single package and interconnecting them.

The large dual in-line package shown in Figure 9-8 contains three monolithic chips. A network of conductors have been formed on the same base that supports the chips. The various conductor pads on the chips are connected to these conductors with fine gold wires that have been bonded in place. The conductors are connected to the two rows of leads along the edge of the package. The lid (or cover) shown immediately below the package, is placed over the opening in the package and soldered into place to provide an air tight (hermetically sealed) unit.

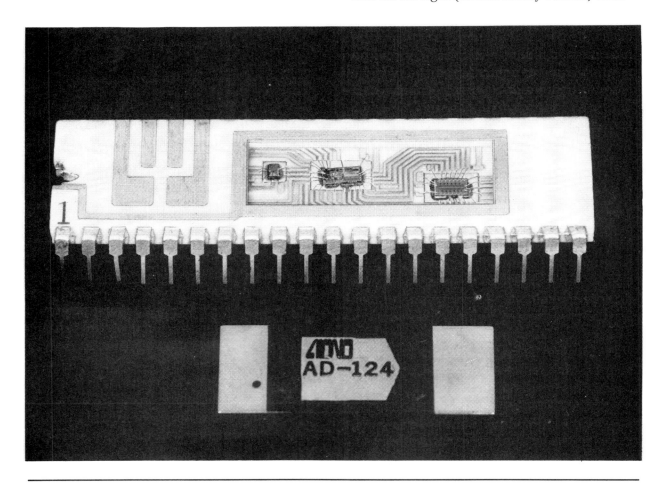

Figure 9-8 A large dual in-line package which contains three monolithic chips. (Courtesy of Precision Monolithics).

Another type of IC package that is widely used, is the flat-pack. The flat-pack is similar to the dual in-line package but it is smaller and much thinner. A typical flat-pack is illustrated in Figure 9-9. Notice that the device is very thin and its leads extend horizontally outward around its edges. The flat-pack can be mounted almost flush with the surface of a printed circuit board and its leads are usually soldered directly to adjacent conductor pads on the circuit board. The flat-pack is used where space is limited. It is often made from metal and ceramic materials and can be used over a wide range of operating temperatures (-55 to $+125$ degrees centigrade). Flat-packs are widely used with monolithic circuits to take full advantage of their extremely small size.

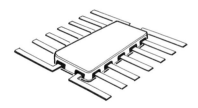

Figure 9-9 A typical flat-pack.

Integrated circuits may also be mounted in metal cans that are similar to the types used to house transistors. A typical metal can is shown in Figure 9-10. The metal can shown has only 8 leads, although larger devices with more leads are available. Metal cans may be used to house monolithic, film-type, or hybrid IC's and their long metal leads make it possible to install them in a variety of ways. Metal cans may be used over a wide temperature range (-55 to $+125$ degrees centigrade) and are therefore suitable for military and space applications.

Hybrid integrated circuits and film circuits are constructed on insulating substrates (usually ceramic materials), which in some cases serve as part of their final package. For example, the substrate that was shown in Figure 9-5 has input and output terminals around its outer edge to which pins can be attached. With the pins attached, the substrate can be plugged into a matching socket or plugged into a printed circuit board and soldered into place. All that is needed to complete the package is a ceramic lid or cover that can be placed on top of the substrate to protect the components from the environment.

Figure 9-10 A typical metal can.

After the integrated circuits are packaged in their respective containers, they are put through a series of tests to be sure that they meet certain electrical specifications. Since the performance of any circuit is affected by changes in temperature, the IC's must be tested over a wide range of operating temperatures to insure that they are suitable for either commercial and industrial applications, or military and space operations.

11. The monolithic IC is produced by using the same basic techniques that are used to produce a bipolar transistor.

    A. True
    B. False

12. As many as 500 IC's can be simultaneously formed on a circular semiconductor _____.

13. The semiconductor material on which the IC's are formed is commonly referred to as a _____.

14. When the monolithic IC's have been formed they are sliced into individual sections which are called _____ or _____.

15. The number of usable IC's obtained in a given process is usually expressed as a percentage of the maximum number possible and is referred to as the _____.

16. The value of a monolithic IC resistor can be adjusted by physically varying its _____ and _____.

17. The two types of capacitors often used in monolithic IC's are referred to as _____ and _____ capacitors.

18. In general, the size of a monolithic resistor or capacitor increases as its value _____.

19. Integrated circuits are classified according to complexity or the number of components they contain and are identified as _____, _____ or _____ circuits.

20. The monolithic IC in Figure 9-2 utilizes _____ isolation as a means of electrically separating components.

21. Monolithic components may also be separated by means of _____ isolation or _____ isolation.

22. Monolithic IC's that utilize bipolar transistor construction techniques are shown in Figure 9-2, are known as _____ IC's.

23. Monolithic IC's that utilize MOSFET's as the principle controlling elements are referred to as _____ IC's.

24. Thin-film IC's are formed by depositing metals and oxides in extremely thin layers on an _____ substrate.

25. With thin-film techniques, it is possible to produce extremely accurate resistance values.

    A. True
    B. False

26. Thin-film techniques can be used to produce diodes and transistors, as well as resistors and capacitors.

    A. True
    B. False

27. The materials used to form thin-film circuits are deposited by using an _____ or _____ process.

28. In thick-film circuits, resistors, capacitors and conductors are formed on an insulating substrate with a _____ process.

29. Hybrid IC's are devices which contain combinations of _____, _____ and _____ circuits.

30. Three integrated circuit packages that are widely used are the _____ package, the _____, and the _____.

Applications Of IC's

Integrated circuits may be placed into two general categories. They can be classified as either digital IC's or as linear IC's. Digital IC's are the most widely used devices. They are simply switching circuits which handle information and they are designed for use in various types of logic circuits and in digital computers. A linear IC provides an output signal that is proportional to the input signal applied to the device. Linear IC's are widely used to provide such functions as amplification and regulation. They are often used in television sets, FM receivers, electronic power supplies, and in various types of communications equipment.

It is impossible to consider all of the possible applications of digital and linear IC's. However, we can examine a few typical examples. First we will examine digital IC's and then typical linear ICs.

Digital IC's

Digital circuits use discrete values (either 0 or 1) to perform 3 general functions. They are the AND, OR, and NOT functions. The 3 functions are performed by logic circuits that are called the AND, OR, and NOT logic gates. These gates or circuit configurations can be combined to make decisions based on digital input information. Thus, the gates are also called digital logic gates.

It is not practical at this time to go into a detailed description of all of the digital ICs that are available. That material will be explained in detail in Heathkit/Zenith courses such as Digital Techniques and Microprocessors.

At this time it is necessary only to define each of the 3 basic functions. This can be accomplished using truth tables that show all of the possible inputs (A and B) and outputs of the AND, OR and NOT logic gates. In a digital gate it is only possible to have an output of either 0 or 1. Zero and 1 are representations for voltage levels. Usually a voltage of approximately 3 volts or higher is considered to be a logic 1 and voltages of 0.7 and below are considered to be logic 0.

AND Gate

The AND gate has an output only when all of its inputs are equal to 1. This is similar to a multiplier function since the only possibilities in a digital circuit are $0 \times 1 = 0$ and $1 \times 1 = 1$.

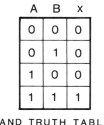

A	B	X
0	0	0
0	1	0
1	0	0
1	1	1

AND TRUTH TABLE

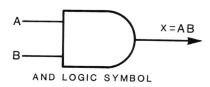

AND LOGIC SYMBOL

$$x = AB$$

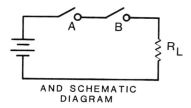

AND SCHEMATIC
DIAGRAM

Figure 9-11 AND gate truth table, logic symbol, and schematic diagram.

The schematic drawing in Figure 9-11 has 2 switches connected in series. Unless both switches are closed, there is no current flow to the output.

OR Gate

The OR gate provides a 1 in the output when any or all of its inputs are 1. This is similar to an adder function except that it doesn't provide for a carry. Thus, $0 + 0 = 0$, $0 + 1 = 1$, and $1 + 1 = 1$. The + symbol doesn't really mean add, it means that A and B are ORed together.

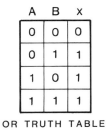

A	B	X
0	0	0
0	1	1
1	0	1
1	1	1

OR TRUTH TABLE

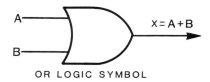

OR LOGIC SYMBOL

$$x = A + B$$

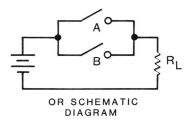

OR SCHEMATIC
DIAGRAM

Figure 9-12 OR gate truth table, logic symbol, and schematic diagram.

The schematic drawing in Figure 9-12 has 2 switches connected in parallel and if either or both switches are closed there is a complete path for current flow.

NOT Gate

The NOT gate provides an output that is always opposite the input. This is called inversion or 180 degrees of phase shift. Thus, the NOT gate is commonly referred to as an inverter. Remember that the common emitter amplifier configuration was the only amplifier configuration capable of inverting the input signal. Therefore, the common emitter configuration is capable of performing the NOT function and when used for as an inverter it is driven between saturation and cutoff. This provides outputs of either Vcc or 0.7 volts.

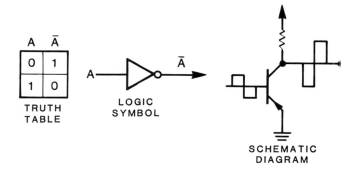

Figure 9-13 NOT gate truth table, logic symbol, and schematic diagram.

Combinational Logic Circuit

This circuit combines the AND, OR, and NOT functions into a single decision making circuit with 3 distinct outputs.

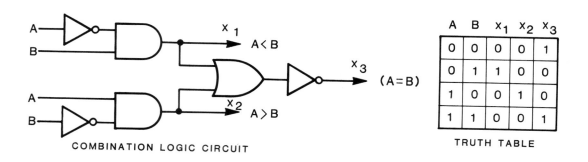

Figure 9-14 Combined logic symbols.

In Figure 9-14 the circuit uses 2 AND gates, 1 OR gate, and 3 NOT gates to provide outputs when A is less than B, A is greater than B, and when A and B are equal. This is sometimes referred to as a choice box in a digital computer program.

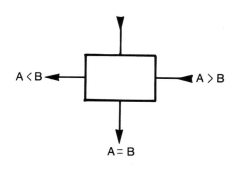

Figure 9-15 Logic symbol for decision making circuit.

Typical ICs

Most of the integrated circuits in use today are digital IC's. These devices are widely used in digital computers and portable electronic calculators to perform various arithmetic and decision making functions. Digital IC's are produced using both bipolar and MOS construction techniques. These circuits can be very simple or extremely complex and are available in the SSI, MSI, and LSI levels.

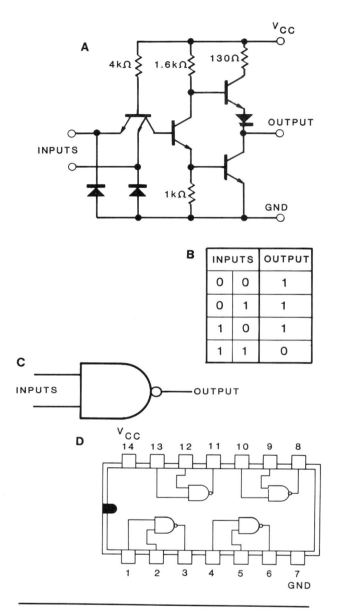

Figure 9-16 A typical TTL digital IC.

A typical digital IC which is formed by using bipolar construction techniques is shown in Figure 9-16. A schematic diagram of the IC is shown in Figure 9-16A. Notice

that only transistors, diodes, and resistors are used in the circuit and since it contains only 11 components, it is classified as an SSI circuit. In this circuit, the transistors are the key elements and because of the unique manner in which they are connected, the circuit is commonly referred to as a transistor-transistor logic (TTL) circuit.

The TTL circuit in Figure 9-16A performs an important logic function. It is capable of comparing two input voltage levels, which must be equal to either 0 volts or approximately 3 volts, and provide an output voltage level (0 or 3 volts) depending on the input combination. The circuit performs what is commonly referred to as the NAND function and the circuit itself is referred to as a NAND gate since it provides a gating or switching action between two voltage levels. The NAND gate, like all digital circuits, is capable of recognizing only two voltage levels (sometimes called logic levels) at each of its inputs. Instead of referring to the specific voltages involved (which can vary with different types of digital circuits), it is common practice to refer to one level as a high logic level (a logic 1) and the other voltage level as a low logic level (a logic 0). A digital circuit can be thought of as a device which responds to various high and low (1 or 0) logic levels and the actual voltages involved can be ignored. The table in Figure 9-16B shows the output levels (1 or 0) produced by the NAND gate when all possible combinations (only 4 are possible) of input levels have been applied. The NAND gate is capable of making a decision based on the combinations of logic levels at its inputs and provide a specific output logic level for each combination.

This is only a basic introduction to digital logic circuits and a complete analysis of digital circuits and techniques are explained in detail in the Heathkit/Zenith Digital Techniques course.

Since many thousands of NAND gates are used in digital computers and other complex digital systems, it is not practical to draw the entire circuit each time it is shown on a schematic. Therefore, the NAND gate is usually represented by the symbol shown in Figure 9-16C. Notice that only the two inputs and the output of the circuit are represented.

It is common practice to construct not one but four of these NAND gate circuits on a single IC chip, using bipolar techniques, and to mount the chip in a single package. Both dual in-line packages and flat-packs are widely used with IC's of this type. The package outline of a typical dual in-line package is shown in Figure 9-16D. This package outline shows how the various NAND gates are internally connected to the package leads. Notice that the IC package has 14 leads (also called pins), which are consecutively numbered in a counterclockwise direction. The package has a notch at one end, which serves as a key to help locate pin number 1. Notice that pins 1 and 2 serve as inputs to one gate and pin 3 provides the output connection for this particular IC. Power is simultaneously applied to all four circuits through pins 7 and 14.

Looking down from the top the pins are always numbered counterclockwise but the function and power requirements of a particular pin may vary from manufacturer to manufacturer. You should always check the manufacturer's data sheet for your particular IC before wiring it into a circuit.

The package outline drawing shown in Figure 9-16D is typical of those provided by IC manufacturers in their specification sheets. They may also provide a schematic of the particular circuit involved and, of course, will always provide the important electrical characteristics of the circuit. In many cases, the circuit designer or engineer is more interested in what the circuit can do and is less interested in how it does it or how it is constructed. Therefore, specification sheets are likely to contain more mechanical and electrical information which pertains to the overall performance of the IC and very little information relating to its internal construction.

A typical example of a digital IC which is formed by using MOS construction techniques is shown in Figure 9-17. This IC utilizes one of the newest and most advanced construction techniques. It contains both P-channel and N-channel enhancement mode MOS field-effect transistors and is commonly referred to as a complementary-symmetry/metal-oxide semiconductor IC or simply a CMOS IC. CMOS circuits containing P- and N-channel MOSFET's are now widely used because they have many advantages over other types of digital circuits. They consume less power than other types of digital IC's and they have good temperature stability. They can operate over a wide range of supply voltages (typically 3 to 15 volts), as compared to TTL circuits which require an accurate 5 volt supply. CMOS circuits also have a high input resistance, which makes it possible to connect a large number of circuit inputs to a single output without loading down the output and disrupting circuit operation. This is an extremely important advantage in digital equipment where thousands of circuits are used. The main disadvantages of CMOS devices are their slow speed of operation and they are more likely to be damaged by an electrolytic charge (static electricity). Additional grounding precautions should be observed when handling CMOS devices.

The circuit shown in Figure 9-17A contains four MOSFET's which are interconnected so

that they can perform a useful logic function. The resulting circuit is referred to as a NOR gate, and like the NAND gate previously mentioned, it is a fundamental building block that is used to construct complex digital circuits. However, the NOR gate responds differently to various combinations of input voltage levels. The output levels (1 or 0) produced by the NOR gate for all possible combinations of input levels are shown in the table in Figure 9-17B.

The NOR gate symbol is shown in Figure 9-17C. This symbol is generally used in place of the actual schematic. Four of these NOR gates are usually formed on one IC chip and mounted in a single IC package. An outline drawing of a typical dual in-line package, which contains four NOR gates, is shown in Figure 9-17D.

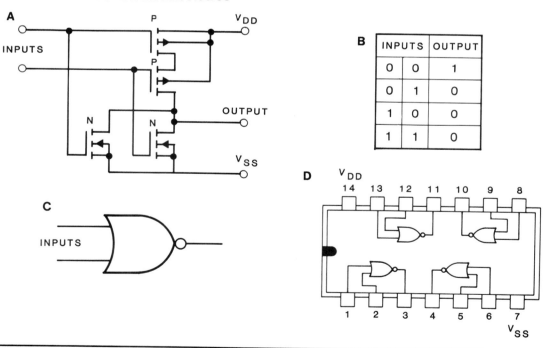

Figure 9-17 A CMOS digital IC.

Both TTL and CMOS circuits can be used to perform NAND or NOR functions or a variety of other logic functions that must be performed in a highly complex digital system. The circuits shown in Figures 9-16 and 9-17 are typical of the SSI circuits used in digital equipment. These circuits may be thought of as basic building blocks which can be used to construct complex digital systems that can perform useful operations.

Linear IC's

Linear circuits provide outputs that are proportional to their inputs. They do not switch between two states like digital circuits. The most popular linear circuits are the types that are designed to amplify DC and AC voltages. In fact, a high performance amplifier circuit, known as an operational amplifier, is widely used in various types of electronic equipment.

The operational amplifier can amplify DC or AC voltages and has an extremely high gain. An operational amplifier can be constructed with discrete components, but it is more commonly produced in IC form and sold as a complete package which is designed to meet certain specifications. However, the operational amplifier is designed so that it can be used in a variety of applications. You can control its gain by using additional external components and it usually has built-in features which make it possible to adjust its operation in a variety of ways.

The basic operational amplifier or op amp is shown in Figure 9-18.

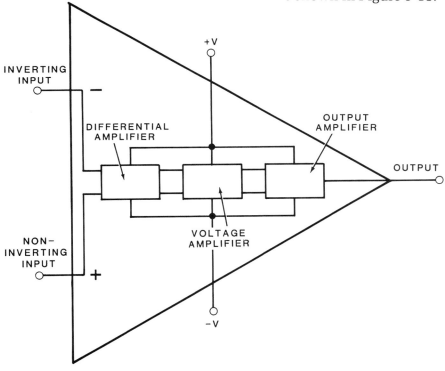

Figure 9-18 Block diagram of a basic op amp.

Note that the standard op amp is really 3 amplifiers contained in the same package. It is a common practice to put 4 op amps into a single chip as you will see when you get to the experiments that are part of this course of instruction. The standard op amp has 2 inputs marked (−) and (+). The negative (−) input is the inverting input and when used it will provide an output that is shifted in phase by 180 degrees. The positive (+) input is the noninverting input and when used the output will be the same phase as the input.

When the op amp is used as an amplifier without feedback it has a DC gain that approaches infinity. Most op amps are considered to have a gain bandwidth product of 1 MHz. Therefore an op amp operated without feedback in the audio range of approximately 20 kHz would have a gain of approximately 50. The term feedback means that part of the output is coupled back to the input to improve stability. This also gives you more control over the circuits operation. This type of feedback is called degenerative feedback because it opposes the input signal and results in a decreased circuit gain.

When an op amp is operated without feedback it is easily saturated because of its high gain. Its output will be either zero when not conducting or within 2 volts of its supply voltage when saturated. Since the op amp has both negative and positive supply voltages the output could swing almost 2 × supply voltage. For example, if the supply voltages were plus and minus 12 volts the

output could swing from + 10 volts to − 10 volts as shown in Figure 9-19.

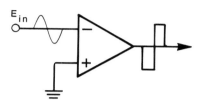

Figure 9-19 Inverting amplifier without feedback.

The circuit shown in Figure 9-19 is an excellent wave shaping circuit. As you can see when a small amplitude sign wave is applied to the input you can obtain a relatively high amplitude square wave output. By routing the square wave through a diode of the proper polarity you can obtain either a positive or negative digital pulse train. The pulse train could then be used to represent logic zeros and ones. Note that the output in Figure 9-19 is inverted, this is because the input is applied to the inverting input of the op amp. If polarity inversion was not desired the inputs could be reversed.

The op amp can also be used as a buffer amplifier. The buffer amplifier is much like the common collector amplifier configuration discussed earlier. You may recall that it was also called an emitter follower. Since the op amp does not have an emitter the circuit is simply called a voltage follower. In other words what ever happens to the input also happens in the output. See Figure 9-20.

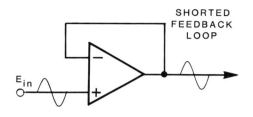

Figure 9-20 Op amp with maximum feedback.

Like the common collector configuration, the circuit shown in Figure 9-20 has a very low output impedance and a voltage gain of slightly less than 1.

The operational amplifier can have a voltage gain by simply replacing the shorted feedback loop with a resistor. The circuit in Figure 9-21 is an example of an op amp with a gain of -10. The minus sign indicates that the output is inverted when compared to the input. The gain of an op amp depends on the ratio of R_F/R_{in}.

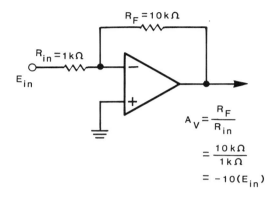

$$A_V = \frac{R_F}{R_{in}}$$

$$= \frac{10\,k\Omega}{1\,k\Omega}$$

$$= -10(E_{in})$$

Figure 9-21 Inverting amplifier with a gain of -10.

This information was presented as an introduction to op amps and some of their uses and characteristics. Op amps, op amp circuits, and applications are discussed in more detail later in your study of the Heathkit/Zenith course Electronic Circuits. A typical operational amplifier circuit is shown in Figure 9-22A. The circuit contains transistors, resistors, and capacitors which are interconnected to form a highly efficient amplifying circuit. The circuit has two inputs and one output as shown. One input is commonly referred to as the (+) or non-inverting input and the other is referred to as the (−) or inverting input. The circuit will amplify either DC or AC signals applied to either (+ or −)input. However, signals applied to the (+) input are not inverted at the output. In other words, as the input voltage goes positive or negative, the output voltage goes positive or negative. When a signal is applied to the (−) input, inversion takes place. In other words, the polarity of the output signal is always opposite to that of the input signal. This unique feature greatly increases the versatility of the operational amplifier (op amp).

When the input voltage is equal to zero, the output voltage should also be equal to zero. However, in practice the output voltage may be offset by a slight amount since component tolerances make it impossible to construct a perfectly balanced circuit. Therefore, two offset null terminals are provided so that the circuit can be appropriately balanced. This is done by simply connecting the opposite ends of a potentiometer of the offset null terminals, but the arm of the potentiometer is connected to circuit ground. The potentiometer may then be adjusted to balance the circuit.

Power is applied to the operational amplifier through terminals V + and V − . The circuit requires a positive voltage source and a negative voltage source. Most operational amplifiers can operate over a reasonably wide range of supply voltages, but these voltages should never exceed the maximum limits set by the manufacturer of the device. Operational amplifiers also consume very little power. Most units have maximum power dissipation ratings of 500 milliwatts or less.

The operational amplifier has an extremely high input resistance, but its output resistance is very low. The device also has an extremely high voltage gain. Many units are guaranteed to amplify an input voltage by at least 15,000 to 20,000 times and some of these units have typical gains that can exceed several hundred thousand.

In most cases, operational amplifiers are not used alone and their full amplifying capabilities are not utilized. Instead, it is common practice to connect external components to the amplifier in a manner which will allow a small portion of the output signal to return to the input and control the overall gain of the circuit. A lower gain is obtained in this manner, but circuit operation becomes more stable and predictable.

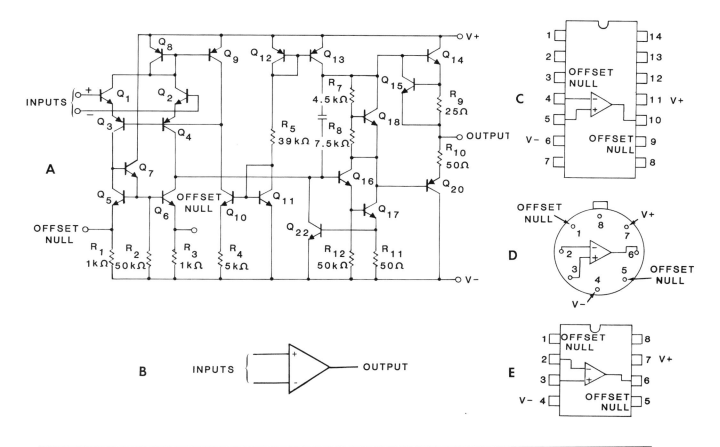

Figure 9-22 An operational amplifier.

The operational amplifier is commonly represented by the symbol shown in Figure 9-22B. Notice that the (+) and (−) inputs are identified in the symbol. The operational amplifier circuit is generally packaged in a variety of ways to suit a broad range of applications. For example, the circuit in Figure 9-22A is available in a dual in-line package (DIP) as shown in Figure 9-22C or in a metal can package as shown in Figure 9-22D. It is even available as a mini DIP as shown in Figure 9-22E.

Operational amplifiers are used in various types of electronic equipment. They are the most important components used in electronic analog computers because their linear characteristics can be used to provide multiplying and summing operations. When used in analog computers, voltages are used to represent (are analogous to) the actual quantities to be multiplied or added. The extremely small operational amplifier IC's are suitable for use in portable electronic equipment where weight and power consumption must be held to a minimum. They are also suitable for use in portable test instruments and in communications equipment.

Operational amplifiers are explained in more detail in the Heathkit/Zenith Operational Amplifiers course and the Electronic Circuits course.

Various types of voltage regulator circuits are also constructed in IC form. These linear devices are used to convert an unregulated DC voltage (obtained through AC rectification) into a regulated DC output voltage which remains essentially constant while supplying a wide range of output currents. These voltage regulator IC's have replaced many of the discrete component regulators that were once widely used. Some IC regulators provide only one fixed output voltage, but other types are available which have adjustable outputs.

Special types of linear IC's are also designed for specific applications. For example, special linear IC's are designed for use in FM receivers where they are used to detect (recover the information in) FM signals. Some IC's are designed for use in solid-state television receivers where they are used to detect, process, and automatically control the chroma (color) signals. Others are used to provide extremely simple operations such as generating a signal, operate a lamp or some other indicator when an FM receiver or TV receiver is properly tuned. Some linear IC's are even used as an interface between digital circuits when digital information must be transmitted over a long transmission line. Such devices are commonly referred to as line drivers and receivers and they are used in digital systems even though they are basically linear devices.

31. Integrated circuits may be broadly classified as either _____ or _____ IC's.

32. The IC shown in Figure 9-16C is commonly referred to as a _____ gate.

33. Like all digital circuits, the circuit in Figure 9-16A responds to only _____ input voltage levels.

34. The voltage levels used in a digital system are commonly referred to as _____ and _____ or _____ and _____ logic levels.

35. Power must be applied to the IC package shown in Figure 9-16D through pins _____ and _____.

36. The digital IC in Figure 9-17A utilizes both P- and N-channel _____.

37. The circuit in Figure 9-17A is commonly called a _____ gate.

38. A CMOS digital IC usually has a higher input _____ than a TTL digital IC.

39. A linear circuit provides an output that is _____ to its input.

40. One of the most popular linear IC's is the _____ amplifier.

41. When an AC signal is applied to the (−) input of the circuit shown in Figure 9-22A, it is _____ when it appears at the output.

42. The IC package outlined in Figure 9-22E is commonly referred to as a _____.

43. Positive and negative voltage sources must be applied to the IC package shown in Figure 9-22C through pins _____ and _____ respectively.

The integrated circuit has revolutionized the electronics industry because it has brought about tremendous reductions in the size and weight of electronic equipment. The use of IC's has also resulted in greater equipment reliability and lower operating expense. IC's are now used extensively throughout the electronics industry and new applications for these devices are constantly being found.

Integrated circuits are constructed in basically four different ways. They may be produced by using monolithic, thin-film, thick-film or hybrid techniques. Monolithic IC's are the most widely used devices. Monolithic circuits are formed within a single piece of semiconductor material known as a substrate. The components most commonly found in these circuits are diodes, transistors, and to a lesser extent resistors, and capacitors.

Monolithic circuits can be formed by using bipolar or MOS construction techniques. Bipolar IC's utilize the same basic technology associated with the construction of bipolar transistors. MOS IC's are produced with the same basic technology that is used to construct MOS field effect transistors.

Thin-film and thick-film circuits are formed by depositing layers of metals and oxides on insulating substrates. The components that are formed in a thin-film circuit are so thin that they have almost no discernable thickness. The components formed in thick-film circuits are much larger and more closely resemble discrete components in appearance. Film techniques are used to construct resistors, capacitors, and interconnecting conductors, but components such as diodes and transistors must be formed separately out of semiconductor materials and attached to the film circuit.

Hybrid circuits are formed by combining monolithic, thin-film, and thick-film construction techniques. The hybrid approach to the construction of IC's make it possible to utilize the best feature found in each type of IC.

Integrated circuits can be packaged in a variety of ways. In most cases the package (no matter how small) is considerably larger than the IC itself. The package protects the IC from various contaminants which could affect its operation and at the same time provides suitable mounting pins or leads which can be easily inserted in a matching socket or receptacle or soldered in place. The DIP, mini DIP, flat-pack and metal can packages are among the types that are most commonly used with the DIP being the most common.

All integrated circuits, no matter how they are constructed, can be classified as either digital or linear devices. Digital IC's contain switching circuits which are capable of performing various logic functions by switching from one discrete value to the other (0 and 1). Linear IC's operate continuously and do not switch from one state to another. Linear circuits provide an output that is proportional to their respective inputs.

The operational amplifier is the most common IC used in linear circuits. It is capable of amplifing both AC and DC voltages and it provides a choice of inverted or non-inverted outputs. The op amp has a high input resistance and a very low output resistance and is capable of very high gains.

CMOS IC's combine an N-channel with a P-channel in a single package and is considered to be a complementary-symmetry amplifier much like the combining of a PNP and NPN bipolar transistor in the same circuit. CMOS devices are available for a variety of circuit applications and have the advantage of small size and low power consumption over both MOS and TTL IC's, but they are also slower in operation and more likely to be damaged during handling.

Unit 10
Optoelectric Devices

Contents

Introduction

In this unit you will examine a group of solid-state components which are capable of converting light energy into electrical energy or electrical energy into light energy. These components are commonly referred to as optoelectronic devices since their operation relies on both optic and electronic principles.

The optoelectronic devices described in this unit are divided into two basic groups. They are classified as light-sensitive devices and light-emitting devices. The most important components found within these two categories are described in detail and the various ways in which these components can be used are considered.

Like other fields, optoelectronics has its own peculiar set of terms, definitions and units of measurement. Although it is not necessary to understand all of the terms that are used, you must at least learn certain portions of the optoelectronics language if you are to fully appreciate and understand the operation of the various optoelectronic components. The important concepts that are involved are therefore developed at the beginning of this unit.

Study this unit carefully. A familiarity with optoelectronic devices is becoming increasingly important since these components are now used extensively and new applications are constantly being found. Also, the optoelectronic theory that you learn in this unit will greatly enhance your knowledge of semiconductor devices.

When you have completed this unit on optoelectronic devices you should be able to:

1. Describe the major characteristics of light.

2. Determine the wave length of any given light frequency.

3. Explain the difference between the radiometric and photometric systems for measuring light.

4. Describe the basic function of a light-sensitive device.

5. Name four photosensitive devices and briefly describe their operation.

6. Describe the basic function of a light-emitting device.

7. Describe the basic principle of operation of the light-emitting diode.

8. Name three advantages that the light-emitting diode has over an incandescent or neon lamp.

9. Explain the basic principles of operation of a liquid-crystal display.

10. Identify at least one advantage of liquid-crystal over other types of displays as well as at least one disadvantage of liquid crystal displays.

11. Name the photosensitive device that has the fastest operating characteristics.

12. Distinguish between the LED and photodiode schematic symbols.

Basic Principles Of Light

In order to understand the operation of optoelectronic devices it is necessary to understand the basic principles of light. You must know what light is and you must be familiar with some of its basic units of measurement. Therefore, we will begin this unit by defining light and considering its various properties. Then we will briefly examine the basic techniques which are used to measure light. This initial discussion will include only the most basic and important terms and definitions.

Characteristics

The term light is used to identify electromagnetic radiation which is visible to the human eye. Basically, light is just one type of electromagnetic radiation and differs from other types such as cosmic rays, gamma rays, X-rays, and radio waves only because of its frequency.

The light spectrum extends from approximately 300 gigahertz* to 300,000,000 gigahertz. It is wedged midway between the high end of the radio frequency (RF) waves, which roughly extend up to 300 gigahertz, and the X-rays, which begin at roughly 300,000,000 gigahertz. Above the X-ray region are the gamma rays and then the cosmic rays.

Like other types of electromagnetic radiation, light propagates (travels) through space and certain types of matter. The movement of light through space can be compared to the movement of radio waves which periodically fluctuate in intensity as they move outward from an antenna or some other radiating body. Therefore, light waves can be measured in terms of wavelength. As with radio waves, the wavelength of light waves is determined by the velocity at which light is moving and the frequency of its fluctuations. This relationship can be summed up with the following mathemetical equation:

$$\lambda = \frac{v}{f^*}$$

Where the Greek letter lambda, λ, represents one complete wavelength, the letter v represents velocity and f represents frequency. This equation states that one complete wavelength is equal to the velocity divided by the frequency.

Light travels at an extremely high velocity. In a vacuum its velocity is 186,000 miles per second or 3×10^{10} (30,000,000,000) centimeters per second. Its velocity is only slightly lower in air but when it passes through certain types of matter, such as glass or water, its velocity is reduced considerably.

When we substitute the velocity of light in a vacuum (in centimeters per second) into the equation given above along with its frequency in hertz, we obtain the wavelength

*One gigahertz, abbreviated GHz, is equal to a frequency of 1,000,000,000 Hertz or 10^9 cycles per second.

in centimeters (cm). Since light has a frequency range which extends from 300 gigahertz (300×10^9 hertz) to 300,000,000 gigahertz (300×10^{15} hertz) its wavelength will vary over a considerable range. The minimum wavelength occurs at the maximum frequency and is equal to:

$$\lambda = \frac{3 \times 10^{10}}{300 \times 10^{15}}$$

$$= 0.0000001 \text{ centimeter}$$

The maximum wavelength occurs at the minimum frequency and is equal to:

$$\lambda = \frac{3 \times 10^{10}}{300 \times 10^9}$$

$$= 0.1 \text{ centimeter}$$

A frequency range of 300 to 300,000,000 gigahertz therefore corresponds to wavelengths which vary from 0.1 to 0.0000001 centimeters. Since the wavelengths associated with the various light frequencies are extremely short, it is common practice to express them in much smaller units. For example, 0.0000001 centimeters (0.0000001×10^{-2} meters) may be expressed as 0.001×10^{-6} meters or 0.001 micrometers (0.001 μm). However, it may also be expressed as simply 0.001 microns. Also, 0.0000001 centimeters is equal to 1×10^{-9} meters or 1 nanometer (1nm). Another unit that is commonly used is the angstrom which is represented by the symbol A. One angstrom equals 1×10^{-10} meters. Therefore, 0.0000001 centimeters

(which is equal to 10×10^{-10} meters) is equal to 10 angstroms.

It is important to understand the relationship between frequency and wavelength when dealing with either radio waves or light waves. These two terms are often used interchangeably in any discussion which concerns the utilization of electromagnetic radiation. You will see both of these terms used as you proceed through this unit.

Within the 300 to 300,000,000 GHz light spectrum only a narrow band of frequencies can actually be detected by the human eye. This narrow band of frequencies appears as various colors such as red, orange, yellow, green, blue and violet. Each color corresponds to a very narrow range of frequencies within the visible region. The entire visible region extends from slightly more than 400,000 GHz to approximately 750,000 GHz. Above the visible region (between 750,000 and 300,000,000 GHz) the light waves cannot be seen. The light waves which fall within this region are referred to as ultraviolet rays. Below the visible region (between 300 and 400,000 GHz) the light waves again cannot be seen. The light waves within this region are commonly referred to as infrared rays. The entire light spectrum is shown in Figure 10-1 so that you can compare the three regions just described.

Although light is assumed to propagate as electromagnetic waves, the wave theory alone cannot completely explain all of the phenomena associated with light. For example, the wave theory may be used to explain why light bends when it flows through water or glass. However, it cannot explain the action that takes place when light strikes certain types of semiconductor materials, and it is this resultant action that forms the basis for much of the optoelectronic theory presented in this unit. In order to explain why and how semiconductor materials are affected by light it is necessary to assume that light has additional characteristics.

To adequately explain the operation of the optoelectronic devices included in this unit it is necessary to consider an additional aspect of light as explained by basic quantum theory. Quantum theory acknowledges that light has wave-like characteristics but it also states that a light wave behaves as if it consisted of many tiny particles. Each of these tiny particles represents a discrete quanta or packet of energy and is called a photon.

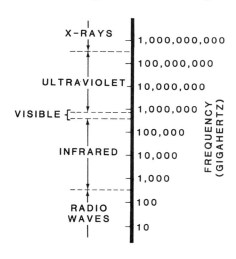

Figure 10-1 The light spectrum.

The photons within a light wave are uncharged particles and their energy content is determined by the frequency and wavelength of the wave. The higher the frequency, the more energy each photon will contain. This means that the light waves at the upper end of the light spectrum possess more energy than the ones at the lower end of the spectrum. This same rule also applies to other types of electromagnetic radiation. For example, X-rays have a higher energy content than light waves while light waves possess more energy than radio waves.

Therefore, light has a dual personality. It propagates through space like radio waves, but it behaves as if it contains many tiny particles. This particle-like property of light will be utilized in this unit to explain the action that takes place in various types of optoelectronic components.

Units of Measurement

In your previous studies you learned how current, voltage, and resistance are related. But even more important was the fact that you learned how to measure these electrical quantities. It was necessary to create suitable units of measurement for each of these quantities so that their precise values could be determined. The same is true when dealing with light. We must know how to accurately measure the various properties of light so that it can be effectively utilized. However, when dealing with light, several problems arise which are not encountered when dealing with the electrical quantities mentioned above. First of all, there is not one but two systems in common use for measuring light. Furthermore, these two systems are similar but not identical and it is often difficult to convert from one system to the other.

One system of light measurement is based on the response of the human eye and is applicable only to that portion of the light spectrum that is visible (approximately 400,000 to 750,000 GHz). This method of describing or quantifying light is referred to as the photometric system. The second system is used to measure both visible and non-visible light. Therefore, it is applicable throughout the entire light spectrum. This method is referred to as the radiometric system.

Since the photometric and radiometric systems are both used in the optoelectronics field, we will briefly examine both of these systems. However, at this time we will consider only the most basic and important terms and units of measurement that are involved in each system. The less important terms will be described later in the unit.

Radiometric System

In the radiometric system the basic units of measurement are the watt and the centimeter and all of the additional units within this system are derived from these. A number of radiometric terms are required to adequately define and measure the various aspects of the light that is emitted from a given source or the light that strikes the sur-face of an object. Furthermore the names of these various radiometric quantities always have the prefix radiant before them. With these thoughts in mind, we will now briefly review the more important radiometric terms and units of measurement.

In radiometric terminology the energy traveling from a light source is considered to be in the form of electromagnetic waves and is referred to as radiant energy. Radiant energy is often represented by the symbol Q_e.

The total amount of radiant energy supplied by a light source per unit time, or in other words, the total rate of flow of the radiant energy is commonly referred to as radiant flux or radiant power. Radiant flux is commonly represented by the symbol Φ_e and is measured in joules per second or watts.

The term radiant intensity may also be used to describe the amount of light produced by a source. However, this term is more specific than the one previously defined and can be used to describe the distribution of radiant energy. Radiant intensity is simply the radiant flux per unit solid angle that is traveling in a given direction. It is represented by the symbol I_e, and it is measured in watts per steradian. The steradian is a dimensionless unit much like the radian. It is simply a unit of solid angular measurement. One steradian of solid angle which has its apex at the center of a sphere that has a radius of 1 meter will subtend an area of 1 square meter on the sphere's surface. Any given sphere will contain a total of 4π steradians.

The two previous terms are useful in describing the radiant energy emitted from a point source of light which radiates in all directions. However, additional terms are needed to describe the radiant energy that is either emitted from or striking a specified surface area. For example, the radiant energy striking a surface per unit area is referred to as irradiance or radiant incidence and is represented by the symbol E_e. It is usually measured in watts per square meter. However, the radiant energy that is emitted from or leaving a surface per unit area is referred to as radiant exitance. Radiant exitance is represented by the symbol M_e and is also measured in watts per square meter.

Another radiometric term that is widely used is radiance. Like the previous term radiant exitance, this term applies to radiant energy that is spread over a specified area. Radiance is defined as the radiant intensity per unit area that is either leaving, passing through, or arriving at a surface in a certain direction. However, in this case the surface area is the area as viewed from the specified direction. Radiance is represented by the symbol L_e and is measured in watts per steradian per square meter. For a wide area light source, the radiance remains constant at all viewing angles with respect to the source. This is because the radiant intensity, which is greatest when the surface area is viewed from directly above, decreases proportionally with the apparent area of the source as the viewing angle (from directly above) increased.

Additional radiometric terms exist but the terms just described are the ones that are most commonly used. It is not necessary to remember all of these terms at this time. Their full meaning cannot be realized until they are used in various practical applications.

Photometric System

The various concepts that were just described using radiometric terminology can also be described using photometric terms. However, the photometric terms always have the prefix luminous before them. Furthermore, the photometric terms are not exactly equivalent to their corresponding radiometric terms. The measurements in photometry are limited to the visible light waves while radiometry deals with the entire light spectrum.

Any photometric light measuring device must respond in much the same manner as the human eye. However, the response of the human eye is slightly different at normal and low light levels. At low light levels the visible light frequencies are shifted to a slightly lower range and the eye is said to have a scotopic response. At normal light levels the eye responds to a slightly higher range of frequencies.

Unless otherwise specified, the various measurements in the photometric system are based on response of the eye. With these thoughts in mind we will examine the various photometric terms and units of measurement which are related to the radiometric terms previously described.

In the photometric system the light energy produced by a given source is referred to as luminous energy. Luminous energy is represented by the symbol Q_v. The amount of luminous energy produced by a source per unit time is referred to as luminous flux or luminous power. Luminous flux is represented by the symbol Φ_v and is measured in lumens. The lumen is considered to be the basic unit of measurement in the photometric system and it may be compared to the watt in the radiometric system. It takes 680 lumens to equal 1 watt of radiant flux, but this is true only for light energy for which the human eye is most sensitive. The conversion is therefore valid only for a specific green light which has a wavelength of exactly 0.555 micrometers (0.555×10^{-6} meters). At any other wavelength, the human eye is less sensitive and there are fewer lumens to each watt. Herein lies the difficulty of converting from radiometric units to photometric units and vice-versa. To make accurate conversions at all wavelengths would require consideration of the spectral response of the human eye and the spectral output of a given light source. Such conversions can be cumbersome and time consuming and will not be considered here.

The luminous flux per unit solid angle traveling in a specific direction from a light source is referred to as luminous intensity. This term may be compared to the radiometric term radiant intensity. Luminous intensity is represented by the symbol I_v and its unit of measurement is the candela. One candela is equal to 1 lumen of luminous flux per steradian.

The luminous flux striking a surface per unit area is referred to as illuminance or illumination and is represented by the symbol E_v. The popular unit of measurement for illuminance is the lux. One lux is equal to 1 lumen per square meter. However, other units of measurement are sometimes used. For example, illuminance may be expressed in lumens per square foot or footcandles. One footcandle equals 10.76 lux.

The luminous flux that is emitted from a unit area of a surface is referred to as luminous exitance. Luminous exitance is represented by the symbol Mv and is measured in lumens per square meter. This term was once known as luminous emittance but this older term is now seldom used.

The last photometric term that we will identify is luminance. Luminance is simply the luminance intensity per unit area that is either leaving, passing through, or arriving at a surface in a specific direction. The surface area is always the apparent area as viewed from the specified direction. Luminance is represented by the symbol Iv and is measured in lumens per steradian per square meter or candela per square meter. One candela per square meter is sometimes called a unit of luminance. Other units are also used which are based on the candela over various units of area but they will not be considered in this brief discussion.

The photometric term luminance may be compared to the radiometric term radiance, although the two terms are not exactly equivalent. Furthermore, the term luminance may be thought of as the photometric equivalent of the term brightness.

As in the radiometric system, additional photometric terms exist. However, the terms just described are the most important ones and they are related to the radiometric terms that were described earlier.

1. Light is considered to be _____ radiation that is visible to the human eye.

2. The light spectrum extends from _____ gigahertz to _____ gigahertz.

3. In terms of wavelength the light spectrum extends from _____ to _____ centimeters.

4. Most of the frequencies within the light spectrum are visible.

   A. True
   B. False

5. The invisible frequencies within the light spectrum are referred to as _____ and _____ rays.

6. Light has wave-like characteristics but it also behaves as if it consisted of many tiny particles known as _____.

7. The _____ system of measuring light is based on the response of the human eye.

8. The _____ system of measuring light is applicable throughout the entire light spectrum.

9. Radiometric terms always use the prefix _____.

10. Photometric terms always use the prefix _____.

11. The radiometric terms _____ _____ and _____ _____ are useful in describing the radiant energy emitted from a point source.

12. The radiometric terms _____ _____, _____ _____ and _____ are used to describe radiant energy that is spread over a specified areas.

13. For each radiometric term there is a corresponding photometric term which describes the same basic concept.

    A. True
    B. False

14. The corresponding radiometric and photometric terms are exactly equivalent.

    A. True
    B. False

Light Sensitive Devices

Most of the optoelectronic devices that are now used are basically light-sensitive devices. In other words they simply respond to changes in light intensity by either changing their internal resistance or by generating a corresponding output voltage. We will now examine some of the most important light-sensitive devices.

Photoconductive Cells

The photoconductive cell is one of the oldest optoelectronic components. Basically it is nothing more than a light-sensitive resistor whose internal resistance changes as the light shining on it changes in intensity. The resistance of the device decreases nonlinearly with an increase in light intensity. That is, the resistance decreases, but the decrease is not exactly proportional to the increase in light.

Photoconductive cells are usually made from light-sensitive materials such as cadmium sulfide (Cd S) or cadmium selenide (Cd Se), although other materials such as lead sulfide and lead telluride have been used. These basic materials may also be doped with other materials such as copper or chlorine to control the exact manner in which the resistance of the device varies with light intensity.

Figure 10-2 shows how a typical photoconductive cell is constructed. A thin layer of light-sensitive material is formed on an insulating substrate which is usually made

from glass or ceramic materials. Then two metal electrodes are deposited on the light-sensitive material as thin layers. The top view (Figure 10-2A) shows that the electrodes do not touch but leave an S-shaped portion of the light-sensitive material exposed. This allows greater contact length but at the same time confines the light-sensitive material to a relatively small area between the electrodes. Two leads are also inserted through the substrate and soldered to the electrodes as shown in the side view in Figure 10-2B. The photoconductive cell is often mounted in a metal or plastic case (not shown) which has a glass window that will allow light to strike the light-sensitive material. Also, the electrodes used with some cells may be arranged in more complicated patterns and may be quite large (1 or more inches in diameter) or relatively small (less than 0.25 inches in diameter).

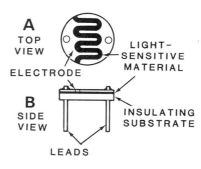

Figure 10-2 Typical photoconductive cell.

Photoconductive cells are more sensitive to light than other types of light-sensitive devices. The resistance of a typical cell might be as high as several hundred megohms when the light striking its surface (its illumination) is zero (complete darkness) and as low as several hundred ohms when the illumination is over 100 lux (approximately 9 footcandles). This represents a tremendous change in resistance for a relatively small change in illumination. This extreme sensitivity makes the photoconductive cell suitable for applications where light levels are low and where the changes in light intensity are small. However, these devices do have certain disadvantages. Their greatest disadvantage is the fact that they respond slowly to changes in illumination. In fact they have the slowest response of all light-sensitive devices. Also, they have a light memory or history effect. In other words, when the light level changes, the cell tends to remember previous illumination. The resistance of the cell at a specific light level is a function of the intensity, the duration of its previous exposure, and the length of time since that exposure.

Most photoconductive cells can withstand relatively high operating voltages. Typical devices will have maximum voltage ratings of 100, 200, or 300 volts DC. However, the maximum power consumption for these devices is relatively low. Maximum power ratings of 30 milliwatts to 300 milliwatts are typical.

The photoconductive cell is often rep-resented by one of the schematic symbols shown in Figure 10-3. The symbol in Figure 10-3A simply consists of a resistor symbol inside of a circle. Two arrows are also used to show that the device is light-sensitive. The symbol in Figure 10-3B is similar but it contains the Greek letter λ (lambda) which is commonly used to represent the wavelength of light.

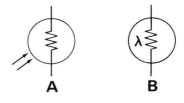

Figure 10-3 Commonly used photoconductive cell symbols.

Photoconductive cells have many applications in electronics. For example, they are often used in devices such as intrusion detectors and automatic door openers where it is necessary to sense the presence or absence of light. However, they may also be used in precision test instruments which can measure the intensity of light. A simple intrusion detector circuit is shown in Figure 10-4. The light source projects a narrow beam of light onto the cell and this causes the cell to exhibit a relatively low resistance. The cell is in series with a sensitive AC relay and its 120 volt AC, 60 Hz power source. The cell allows sufficient current to flow through the circuit and energize the relay. When an intruder breaks the light beam, the cell's resistance increases considerably and the relay is deactivated. At this time the relay contacts close and apply power (from a separate DC source) to an alarm, which sounds a warning. A relay is used because it is capable of controlling the relatively high current that is needed to operate the alarm. When a large relay is used, the photoconductive cell (because of its low power or current rating) may not be able to directly control the relay. In such a case the photoconductive cell is used to control a suitable amplifier circuit which in turn generates enough current to drive the relay.

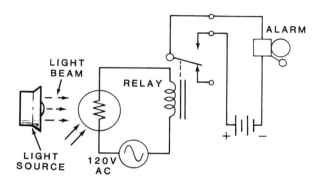

Figure 10-4 A basic intrusion detector circuit.

Since the photoconductive cell is constructed from a bulk material and does not have a PN junction, it is a bidirectional device. In other words it exhibits the same resistance in either direction and may therefore be used to control either DC or AC. Due to its bulk construction, the photoconductive cell is often referred to as a bulk photoconductor. However, you may also see it referred to as a photoresistive cell or simply a photocell.

Photovoltaic Cells

The photovoltaic cell is a device which directly converts light energy into electrical energy. When exposed to light this device generates a voltage across its terminals and this voltage increases as the light increases in intensity. The photovoltaic cell has been used for a number of years in various military and space applications. It is commonly used aboard satellites and spacecraft to convert solar energy into electrical power which can be used to operate various types of electronic equipment. Since most of its applications generally involve the conversion of solar energy into electrical energy, this device is commonly referred to as a solar cell.

The photovoltaic cell is basically a junction device which is made from semiconductor materials. Although many different semiconductor materials have been used, the device is usually made from either silicon or selenium with silicon being the preferred material. The basic structure of a silicon photovoltaic cell is shown in Figure 10-5. Notice that the device has a P-type layer and an N-type layer which form a PN junction and a metal backplate or support is placed against the N-type layer. Also, a metal ring is attached to the outer edge of the P-type layer. These pieces of metal serve as electrical contacts to which the two external leads may be attached.

The photovoltaic cell is designed to have a relatively large surface area which can collect as much light as possible. The cell is constructed so that light must strike the top semiconductor layer within the metal ring as shown in the top view in Figure 10-5A. The device shown has a P-type upper layer but in some cases these cells are designed so that the N-type layer is on top and the P-type layer is on the bottom.

Since the photovoltaic cell has a PN junction, a depletion region (an area void of majority carriers) forms in the vicinity of the junction. If the cell was forward-biased like a conventional PN junction diode, the free electrons and holes in the device would be forced to combine at the junction and forward current would flow. However, the photovoltaic cell is not used in this manner. Instead of responding to an external voltage, the device actually generates a voltage in response to light energy which strikes its surface.

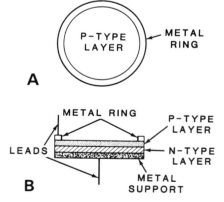

Figure 10-5 A basic silicon photovoltaic cell.
A. Topview   B. Sideview

In order to generate a voltage, the top layer of the photovoltaic cell must be exposed to light. The light energy striking the cell consists of many tiny particles called photons. These photons are actually absorbed at various depths into the semiconductor material depending on their wavelengths and energy. If a particular photon has sufficient energy when it enters the semiconductor material, it can impart much of its energy to an atom within the material.

When sufficient energy is added to the atom, a valence electron may be knocked out of its orbit and become a free electron. This will leave the atom positively charged and a hole will be left behind at the valence site. In other words an impinging photon can produce an electron-hole pair. This electron and hole pair can drift through the semiconductor material. Additional electron-hole pairs are also produced by other photons which penetrate to various depths within the material.

Some of the free electrons and holes generated by the light energy are produced within the depletion region while others are generated outside of the region but are drawn into it. The free electrons in the region are swept from the P-type to the N-type material and the holes are drawn in the opposite direction. The electrons and holes flowing in this manner produce a small voltage across the PN junction, and if a load resistance is connected across the cell's leads, this internal voltage will cause a small current to flow through the load. This current will flow from the N-type material, through the load and back to the P-type material thus making the N and P regions act like the negative and positive terminals of a battery.

All of the photons striking the photovoltaic cell do not create electron-hole pairs and many of the electrons and holes which separate to form pairs eventually recombine. The cell is therefore an inefficient device as far as converting light energy into electrical power. When this efficiency is expressed in terms of electrical power output compared to the total power contained in the input light energy, most cells will have efficiencies that range from 3 percent up to a maximum of 15 percent.

As you might expect, the output voltage produced by a photovoltaic cell is quite low. These devices usually require high light levels in order to provide useful output power. Typical applications require an illumination of at least 500 to 1000 footcandles. At 2000 footcandles, the average open-circuit (no load connected) output voltage of a typical cell is approximately 0.45 volts. When loaded, a typical cell may provide as much as 50 or 60 milliamperes of output load current. However, by connecting a large number of cells in series or parallel, any desired voltage rating or current capability can be obtained.

When used on spacecraft or satellites many photovoltaic cells are connected together, as explained above, to obtain sufficient power to operate electronic equipment or charge batteries. However, these devices are also used as individual components in various types of test instruments and equipment. For example, they are used in portable photographic light meters (which do not require batteries for operation). They are also used in movie projectors to detect a light beam which is projected through the film. The light beam is modulated (controlled) by a pattern or sound track that is printed near the edge of the film. In this way, the intensity of the light beam is made to vary according to the sounds (voice and music) that occured. The photovoltaic cell simply responds to the light fluctuations and produces a corresponding output voltage which can be further amplified and used to drive a loudspeaker which will convert the electrical energy back into sound. This application is shown in Figure 10-6.

A schematic symbol that is commonly used to represent the photovoltaic cell is shown in Figure 10-7. This symbol indicates that the device is equivalent to a one-cell voltage source and the positive terminal of the device is identified by a plus (+) sign.

Figure 10-7 Commonly used symbol for a photovoltaic cell.

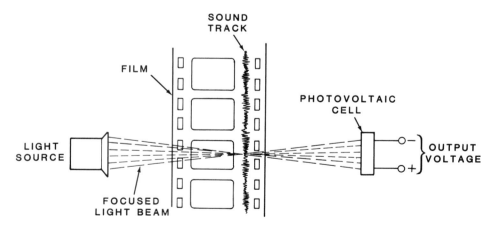

Figure 10-6 A photovoltaic cell used in a movie projector sound reproducing system.

Photodiodes

The photodiode is another light-sensitive device which utilizes a PN junction. It is constructed in a manner similar to the photovoltaic cell just described, but it is used in basically the same way as the photoconductive cell described earlier. In other words it is used essentially as a light-variable resistor.

The photodiode is a semiconductor device (usually made from silicon) and may be constructed in basically two ways. One type of photodiode utilizes a simple PN junction as shown in Figure 10-8A. A P-type region is

diffused into an N-type substrate as shown. This diffusion takes place through a round window that is etched into a silicon dioxide layer that is formed on top of the N-type substrate. Then a metal ring or window is formed over the silicon dioxide layer (through an evaporation process) as shown. This window makes electrical contact with the P-type region and serves as an electrode to which an external lead can be attached. However, the window also accurately controls the area that will receive or respond to light. A metal base is then formed on the bottom N-type layer. This metal layer serves as a second electrode to which another lead is attached.

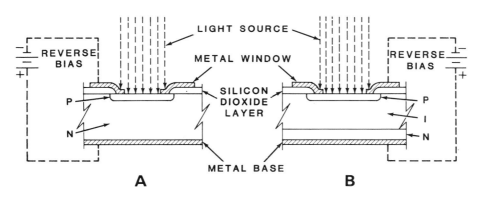

Figure 10-8 Basic construction of typical photodiodes.

The PN junction photodiode shown in Figure 10-8A operates on the same basic principle as the photovoltaic cell previously described. In fact, the photodiode may be used in the same manner as a photovoltaic cell. When used as a photovoltaic cell, the device is said to be operating in the photovoltaic mode and it will generate an output voltage (across its electrodes) that varies with the intensity of the light striking its top P-type layer. However, the photodiode is most commonly subjected to a reverse bias voltage as shown in Figure 10-8A. In other words its P-type region is made negative with respect to its N-type region. Under these conditions a wide depletion region forms around the PN junction. When photons enter this region to create electron-hole pairs, the separated electrons and holes are pulled in opposite directions because of the influence of the charges that exist on each side of the junction and the applied reverse bias. The electrons are drawn toward the positive side of the bias source (the N-type region) and the holes are attracted toward the negative side of the bias voltage (the P-type region). The separated electrons and holes support a small current flow in the reverse direction through the photodiode. As the light intensity increases, more photons produce more electron-hole pairs. This further increases the conductivity of the photodiode resulting in a proportionally higher current. When a photodiode is used in this manner, it is said to be operating in the photoconductive or photocurrent mode.

The photodiode may also be constructed as shown in Figure 10-8B. This type of photo-

diode is similar to the type just described but there is one important difference. This device has an intrinsic (I) layer between its P and N regions and is commonly referred to as a PIN photodiode. The intrinsic layer has a very high resistance (a low conductivity) because it contains very few impurities. A depletion region will extend further into this I region than it would in a heavily doped semiconductor. The addition of the I layer results in a much wider depletion region for a given reverse bias voltage. This wider depletion area makes the PIN photodiode respond better to the lower light frequencies (longer wavelengths). The lower frequency photons have less energy content and tend to penetrate deeper into the diode's structure before producing electron-hole pairs and in many cases it does not produce any electron-hole pairs. The wider depletion region in the PIN photodiode increases the chance that electron-hole pairs will be produced. The PIN photodiode is therefore more efficient over a wider range of light frequencies. The PIN device also has a lower internal capacitance due to the wide I region which acts like a wide dielectric between the P and N regions. This lower internal capacitance allows the device to respond faster to changes in light intensity. The wide depletion region also allows this device to provide a more linear change in reverse current for a given change in light intensity.

PN junction and PIN photodiodes are often mounted on an insulative platform or substrate and sealed within a metal case as shown in Figure 10-9. A glass window is provided at the top of the case, as shown, to allow light to enter and strike the photodiode. The two leads extend through the insulative base at the bottom of the case and are internally bonded (with fine wires) to the photodiode's electrodes.

The performance of a photodiode can be expressed in terms of its quantum efficiency. The quantum efficiency of a particular device is simply the ratio of the electrons produced for each photon that strikes the diode. Quantum efficiency can be expressed mathematically as:

$$\text{quantum efficiency} = \frac{\text{electrons}}{\text{photons}}$$

Figure 10-9 A typical photodiode package.

In the ideal case, one electron should be produced for each photon that strikes the diode thus giving the ideal diode a quantum efficiency of 1. However, the quantum efficiency of a typical photodiode will be lower than 1 and it will vary depending on the wavelength of the radiant energy.

The photo below illustrates the photodiode's physical size.

Two photodiodes are mounted in this 10-lead metal can which has a glass window on top. With this package style, more leads are available than are actually needed.
(Courtesy of Integrated Photomatrix Inc.)

The quantum efficiency of a typical PIN photodiode is plotted graphically in Figure 10-10. The curve in this figure shows how quantum efficiency varies as wavelength varies. Notice that the quantum efficiency is approximately 0.3 (0.3 electrons per photon) at a wavelength of approximately 4,000 angstroms and increases to a peak of approximately 0.8 at 8,000 angstroms. Then it drops off to a value of 0.2 at 10,000 angstroms. This quantum efficiency curve effectively shows how well the photodiode responds to the various wavelengths. This curve shows that the diode's spectral response for all practical purposes extends from approximately 4,000 to 10,000 angstroms.

The performance of a photodiode can also be expressed in terms of its responsivity. The responsivity of a photodiode is simply a measure of how much output (reverse) current is obtainable for a given light energy input. It is expressed as a ratio of output current (called photocurrent) in microamperes to the input radiant energy or irradiance which is measured in milliwatts per square centimeter (mW/cm^2). Expressed mathematically, responsivity is equal to:

$$\text{responsivity} = \frac{\mu A}{mW/cm^2}$$

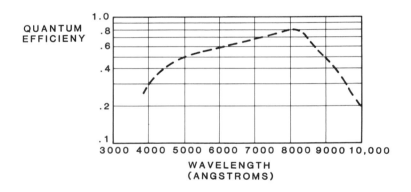

Figure 10-10 Spectral response for a typical PIN photodiode.

The responsivity of a photodiode also varies with the wavelength of the radiant energy striking the device and reaches a peak value when the quantum efficiency of the device is near its highest value. A typical photodiode will have a maximum responsivity of 1.4 microamperes per milliwatt per square centimeter.

The electrical characteristics of a typical PIN photodiode are graphically shown in Figure 10-11. The curves in this figure show the relationship between the voltage across the photodiode and the current flowing through the device when it is used in the photoconductive mode which is the operating mode most widely used. The curves were plotted when the photodiode was exposed to radiant power at a wavelength of 9000 angstroms (0.9 microns) which is well within the diode's spectral response. Notice that the diode's photocurrent is plotted vertically while the reverse voltage applied to the diode is plotted horizontally. The lowest curve shows that when the diode is exposed to input radiant power of 50 microwatts its photocurrent remains almost constant (varies from approximately 15 to 18 microamperes) as its reverse voltage varies from 0 to 30 volts. The next higher curve shows that when the input light power is doubled and the photocurrent almost doubles, but again remains essentially constant (slightly more than 30 microamperes) as reverse voltage varies from 0 to 30 volts. The two upper curves show that photocurrent continues to increase in proportion to the input light power even though the reverse voltage is varied. The curves in Figure 10-11 effec-

tively show that the photodiode produces an output current that is relatively constant and this current is determined primarily by the input light power and to a lesser degree by the diode's reverse voltage. In most applications, the diode's reverse voltage is held constant and its output photocurrent is allowed to vary in proportion to the input light power thus allowing the device to operate in the photoconductive mode.

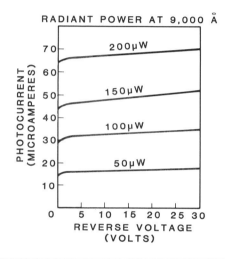

Figure 10-11 Typical photodiode electrical characteristics.

When the input radiant power is reduced to zero or when the photodiode is in complete darkness, the diode's photocurrent will drop to an extremely low value but not quite to zero. This very low current is simply a reverse leakage current which occurs in any type of PN junction device. This very low leakage current is referred to as the photodiode's dark current. The dark current is too small to be plotted in Figure 10-11 since it will generally be in the nanoampere (nano = 10^{-9}) range for most photodiodes. Dark current increases only slightly with an increase in reverse voltage.

Photodiodes have an important advantage over the photoconductive devices described earlier. A photodiode can respond much faster to changes in light intensity. In fact, the photodiode operates faster than any other type of photosensitive device. It is therefore useful in those applications where light fluctuates or changes intensity at a rapid rate. The major disadvantage with the photodiode is that its output photocurrent is relatively low when compared to other photoconductive devices.

Photodiodes and PIN photodiodes are both commonly represented by the same schematic symbol and several symbols have been used to represent these devices. A commonly used symbol is shown in Figure 10-12A. Notice that a conventional diode symbol is used with two arrows. The arrows point toward the diode to show that it responds to light. Figure 10-12B shows a properly biased photodiode. A load resistor (R_L) is also connected in series with the diode. The load resistor simply represents any resistive load which might be controlled by the photodiode as it varies its conductivity in accordance with input light intensity. The changes in the diode's conduction will cause the photocurrent in the circuit to vary.

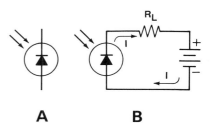

Figure 10-12 A photodiode symbol and a properly biased photodiode.

Phototransistors

The phototransistor is also a PN junction device. However, it has two junctions instead of one like the photodiode just described. The phototransistor is constructed in a manner similar to an ordinary transistor, but this device is used in basically the same way as a photodiode.

The phototransistor is often constructed as shown in Figure 10-13. The process begins by taking an N-type substrate (usually silicon), which ultimately serves as the transistor's collector, and diffusing into this substrate a P-type region which serves as the base. Then an N-type region is diffused into the P-type region to form the emitter. The phototransistor therefore resembles a standard NPN bipolar transistor in appearance. The device is often packaged much like the photodiode shown in Figure 10-9. However, in the case of the phototransistor, three leads are generally provided which connect to the emitter, base, and collector regions of the device. Also, the phototransistor is physically mounted under a transparent window so that light can strike its upper surface as shown in Figure 10-13.

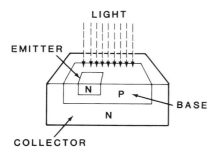

Figure 10-13 The construction of a typical phototransistor.

The operation of a phototransistor is easier to understand if it is represented by the equivalent circuit shown in Figure 10-14. Notice that the circuit shown contains a photodiode which is connected across the base and collector of a conventional NPN bipolar transistor. If the equivalent circuit is biased by an external voltage source as shown, current will flow into the emitter lead of the circuit and out of the collector lead. The amount of current flowing through the circuit is controlled by the transistor in the equivalent circuit. This transistor conducts more or less depending on the conduction of the photodiode which in turn conducts more or less as the light striking it increases or decreases in intensity. If light intensity increases, the diode conducts more photocurrent (its resistance decreases) thus allowing more emitter-to-base current (commonly referred to as base current) to flow through the transistor. This increase in base current is relatively small but due to the transistor's amplifying ability this small

base current can be used to control the much larger emitter-to-collector current (also called collector current) flowing through this device. The increase in input light intensity causes a substantial increase in collector current. A decrease in light intensity would correspondingly cause a decrease in collector current.

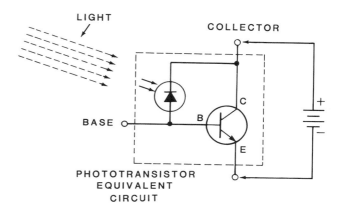

Figure 10-14 Equivalent circuit for a phototransistor.

Although the phototransistor has a base lead as well as emitter and collector leads, the base lead is used in very few applications. However, when the base is used, it is simply subjected to a bias voltage which will set the transistor's collector current to a specific value under a given set of conditions. In other words, the base may be used to adjust the phototransistor's operating point. In most applications only the emitter and collector leads are used and the device is considered to have only two terminals.

The electrical characteristics of a typical phototransistor are graphically shown in Figure 10-15. The collector-to-emitter (collector) current is plotted vertically and the collector-to-emitter (collector) voltage is plotted horizontally. Each curve shown is plotted for a specific amount of irradiance or the amount of light striking the phototransistor per unit area. The lowest curve shows the relationship between collector current and collector voltage for an irradiance of 1 milliwatt per square centimeter. Notice that the collector current quickly jumps from 0 to just slightly less than 1 milliampere and then increases to just slightly more than 1 milliampere as the collector voltage is varied from 0 to more than 16 volts. The collector current, although not constant, varies only a small amount over a wide range of collector voltages. The remaining curves are plotted for higher irradiance values and they effectively show that for each higher value the collector current is proportionally higher and increases only slightly as the collector voltage increases. The transistor's dark current is not shown since it is very low. As with the photodiode previously described, this current increases only slightly with an increase in operating voltage.

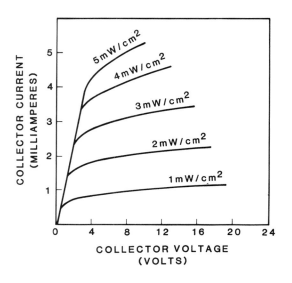

Figure 10-15 Typical phototransistor electrical characteristics.

Figure 10-15 shows that the phototransistor, like the photodiode previously described, provides an output current that is essentially controlled by the intensity of the light striking its surface and to a lesser degree by its operating voltage. The phototransistor is used much like the photodiode by setting its collector voltage to a specific value and allowing the device to control its collector current in accordance with the changes in light intensity.

An important difference between the phototransistor and the photodiode is in the amount of current that each device can handle. The phototransistor can produce much higher output currents than a photodiode for a given light intensity because the phototransistor has a built-in amplifying ability. The phototransistor's higher sensitivity makes it useful for a wider range of applications than a photodiode. Unfortunately this higher sensitivity is offset by one important disadvantage. The phototransistor does not respond as quickly to changes in light intensity and therefore is not suitable for applications where an extremely fast response is required. Like other types of photosensitive devices the phototransistor is used in conjunction with a light source to perform many useful functions. It can be used in place of photoconductive cells and photodiodes in many applications and can provide an improvement in operation. Phototransistors are widely used in such applications as tachometers, photographic exposure controls, smoke and flame detectors, object counting, and mechanical positioning and moving systems.

A phototransistor is often represented by the symbol shown in Figure 10-16A and it is usually biased as shown in Figure 10-16B.

As shown, the phototransistor is used to control the current flowing through a load much like the photodiode shown in Figure 10-12B.

The phototransistor may also be used as a photodiode by using its collector and base leads and leaving its emitter open. When used in this manner, the PN junction between the collector and base region serves as a photodiode. However, when used as a photodiode, the device can handle only a relatively small current, although it will operate at a faster rate of speed.

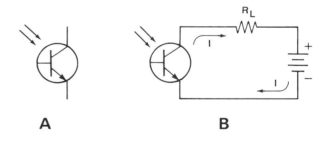

Figure 10-16 A phototransistor symbol and a properly biased phototransistor.

A phototransistor may be interconnected with an ordinary bipolar transistor in a way that allows the phototransistor to control the operation of the bipolar transistor. This type of arrangement is referred to as a photodarlington circuit and it can be simultaneously formed and packaged in a single container. The schematic symbol for a photodarlington arrangement is shown in Figure 10-17 along with the necessary external circuitry. The phototransistor responds to the input light intensity and conducts accordingly. As it conducts (more or less) it controls the base current of the bipolar transistor which in turn controls the current through the load and the battery. Such an arrangement offers a tremendous increase in sensitivity since the gain of the phototransistor is effectively multiplied by the gain of the bipolar transistor, thus producing a relatively high output current. However, the higher gain is obtained by sacrificing speed. Although it is more sensitive than the photodiode or phototransistor, it has a slower response to changes in light intensity.

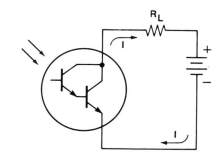

Figure 10-17 A properly biased photodarlington circuit.

15. The photoconductive cell is basically a light-sensitive _____.

16. The resistance of a photoconductive cell decreases as light intensity _____.

17. The photovoltaic cell directly converts light into _____ energy.

18. The photovoltaic cell is sometimes referred to as a _____ cell.

19. Photovoltaic cells are often connected in _____ to produce a higher output voltage.

20. The photodiode may operate in either the _____ or _____ mode.

21. A photodiode which has an intrinsic layer between its P and N regions is referred to as a _____ photodiode.

22. The number of electrons produced for each photon which strikes a photodiode is referred to as the photodiode's _____ _____.

23. The responsivity of a photodiode is a measure of how much reverse _____ is obtainable for a given light energy input.

24. The term dark current refers to the amount of reverse leakage current flowing through a photodiode when the input radiant power is _____.

25. The phototransistor provides a higher output _____ than the photodiode.

26. The phototransistor's _____ and _____ leads are used in most applications.

27. When a phototransistor is connected to a bipolar transistor to obtain greater sensitivity the arrangement is referred to as a _____ circuit.

28. The fastest photosensitive device is the photo-_____.

Light-Emitting Devices

Light-emitting devices are components which produce light when they are subjected to an electrical current or voltage. In other words, they simply convert electrical energy into light energy. For many years incandescent neon lamps were the most popular sources of light in electrical applications. Incandescent lamps use metal filaments which are placed inside a glass bulb and air is drawn out of the bulb to produce a vacuum. Current flows through the filament causing it to heat up and produce light. The neon lamp utilizes two electrodes which are placed within a neon gas filled bulb and current is made to flow from one electrode to the other. When the current flows through the gas, the gas ionizes and emits light.

The incandescent lamp produces a considerable amount of light, but its life expectancy is relatively short. A typical lamp might last as long as 5000 hours. In addition to having a short life, the incandescent lamp responds slowly to changes in input electrical power. The incandescent lamp was (and still is) suitable for use as an indicator or for providing illumination, but due to its slow response time it will not faithfully vary its light intensity in accordance with rapidly charging alternating currents. The incandescent lamp therefore cannot be effectively used to convert high frequency electrical signals (much above the audio range) into light energy suitable for transmission through space. The light energy produced by the incandescent lamp is not useful for carrying information which could subsequently be recovered or converted back into an electrical signal by a suitable light sensitive device.

The neon lamp has a longer life expectancy (typically 10,000 hours) than an incandescent lamp and somewhat faster response to changes in input current. However, its output light intensity is much lower than that of an incandescent device. The neon lamp has been used for many years as an indicator or warning light and in certain applications to transmit low frequency AC signals for information in the form of light over very short distances. The neon lamp cannot be used simply for the purpose of providing illumination.

With all of their shortcomings, incandescent and neon lamps were used for many years because nothing better was available. However, in recent years a new type of light-emitting device was developed which has revolutionized the optoelectronics field. This newer device is a solid-state component, and it is physically stronger than the glass encased incandescent and neon devices. Like all semiconductor devices, it has an unlimited life expectancy. This new light-emitting device is referred to as a light-emitting diode or LED. Since it is such an important solid-state component we will examine the operation and construction of an LED in detail. Then we will see how it is used in various applications.

LED Operation

We have seen how light energy (photons) striking a PN junction diode can impart enough energy to the atoms within the device to produce electron-hole pairs. When the diode is reverse-biased these separated electrons and holes are swept across the diode's junction and support a small current through the device. However, the exact opposite is also possible. A PN junction diode can also emit light in response to an electric current. In this case, light energy (photons) is produced because electrons and holes are forced to recombine. When an electron and hole recombine, energy may be released in the form of a photon. The frequency (or wavelength) of the photons emitted in this manner is determined by the type of semiconductor material used in the construction of the diode.

The LED utilizes the principle just described. It is simply a PN junction diode that emits light through the recombination of electrons and holes when current is forced through its junction. The manner in which this occurs is illustrated in Figure 10-18. As shown in this figure, the LED must be forward-biased so that the negative terminal of the battery will inject electrons into the N-type layer (the cathode) and these electrons will move toward the junction. Corresponding holes will appear at the P-type or anode end of the diode (actually caused by the movement of electrons) and they appear to move toward the junction. The electrons and holes merge toward the junction where

they may combine. If an electron possesses sufficient energy when it fills a hole, it can produce a photon of light energy. Many such combinations can result in a substantial amount of light (many photons) being radiated from the device in various directions.

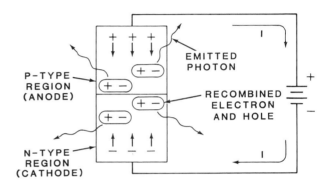

Figure 10-18 Basic operation of a light emitting diode.

At this time, you are probably wondering why the LED emits light and an ordinary diode does not. This is simply because most ordinary diodes are made from silicon and silicon is an opaque or impenetrable material as far as light energy is concerned. Any photons that are produced in an ordinary diode simply cannot escape. LED's are made from semiconductor materials that are semitransparent to light energy. Therefore, in an LED some of the light energy produced can escape from the device.

LED Construction

Many LED's are made of gallium arsenide (GaAs). The LED's made from this material emit light most efficiently at a wavelength of approximately 9000 angstroms which is in the infrared region of the light spectrum and it is not visible to the human eye. Other materials are also used such as gallium-arsenide phosphide (GaAsP) which emits a visible red light at approximately 6600 angstroms and gallium phosphide (GaP) which produces a visible green light at approximately 5600 angstroms. The GaAsP device also offers a relatively wide range of possible output wavelengths by adjusting the amount of phosphide in the device. By adjusting the percentage of phosphide, the LED can be made to emit light at any wavelength between approximately 5500 angstroms to 9100 angstroms.

Although Figure 10-18 helps to illustrate the operation of an LED, it does not show how the device is constructed. The construction of a typical GaAsP LED is shown in Figure 10-20. Figure 10-20A shows a cross-section of the device and Figure 10-20B shows the entire LED chip. The construction begins with a gallium arsenide (GaAs) substrate. On this substrate an epitaxial layer of gallium arsenide phosphide (GaAsP) is grown, however the concentration of gallium phosphide (GaP) in this layer is gradually increased from zero to the desired level. A gradual increase is required so that the crystalline structure of the substrate is not disturbed. During this growth period, an N-type impurity is added to make the epitaxial layer an N-type material. The grown layer is then coated with a special insulative material and a window is etched into this insulator. A P-type impurity is then diffused through the window into the epitaxial layer and the PN junction is formed. The P-type layer is made very thin so that the photons generated at or near the PN junction will have only a short distance to travel through the P-type layer and escape as shown in Figure 10-20A.

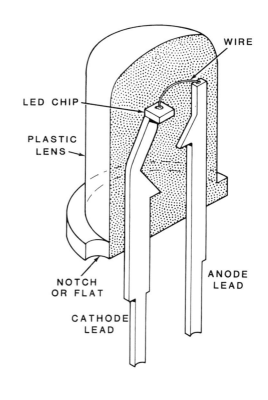

Figure 10-19 A typical LED package.

The construction of the GaAsP LED is completed by attaching electrical contacts to the P-type region and the bottom of the substrate. The upper contact has a number of fingers extending outward so that current will be distributed evenly through the device when a forward bias voltage is applied across the contacts.

Once the LED is formed it must be mounted in a suitable package. Several types of packages are commonly used but all must fulfill one important requirement. All packages must be designed to optimize the emission of light from the LED. This factor is very important because the LED emits only a small amount of light. Therefore most packages contain a lens system which gathers and effectively magnifies the light produced by the LED. Also, various package shapes are used to obtain variations in the width of the emitted light beam or variations in the permissible viewing angle.

The LED package shown in Figure 10-19 is installed by pushing the lens through a suitable hole in a chassis or special bezel and snapping it into place. The leads are then soldered in place. Most LEDs have a flat edge on the body to mark the position of the cathode (negative) lead and the cathode lead is usually longer than the anode (positive) lead. Also when the LED is examined in good light, you may be able to see the inside elements. The cathode element will be the larger of the two elements

A typical LED package is shown in Figure 10-19. As shown, the package body and lens are one piece and are molded from plastic.

The cathode and anode leads are inserted through the plastic case and extend up into the dome shaped top which serves as the lens. The bottom contact of the LED chip is attached directly to the cathode lead and the upper contact is connected to the anode lead by a thin wire which is bonded in place. The placement of the LED chip in the case is critical since the case serves as a lens which conducts light away from the LED and it also serves as a magnifier. In some cases the plastic lens will contain fine particles which help to diffuse the light or the entire case may be dyed or tinted with a color that enhances the natural light color (red lens for red light) emitted by the LED.

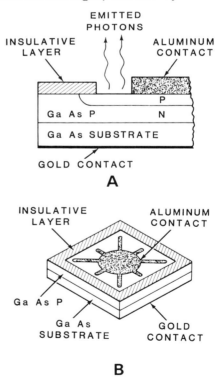

Figure 10-20 Basic construction of an LED.

Typical light-emitting diodes (LED's). Most LED's produce red light, but special devices to generate green or yellow lights are also available.

(Courtesy of Hewlett-Packard)

LED Characteristics

The relationship between forward current and voltage in a typical GaAsP LED is shown graphically in Figure 10-21. Notice that the forward bias must be increased to approximately 1.2 volts before any appreciable forward current flows. Then current increases rapidly for a continued increase in forward voltage. This graph effectively shows that once the LED conducts, its current can vary over a considerable range up to its maximum value while the voltage across the LED remains essentially constant at approximately 1.6 volts. Most LED's exhibit a similar current-voltage relationship.

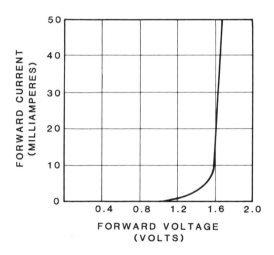

Figure 10-21 Current-Voltage characteristics of a typical GaAsP LED.

Figure 10-22 shows the relationship between the forward current and the total output radiant power produced by a typical GaAsP LED. The power is expressed only in relative terms (as a percentage). The important point to note in Figure 10-22 is that the output radiant power increases linearly with the forward current.

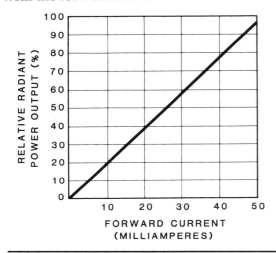

Figure 10-22 Output radiant power versus forward current for a typical GaAsP LED.

The spectral response of a typical GaAsP LED is shown in Figure 10-23. This graph shows the relationship between output radiant power and the wavelength of the radiated energy. The output power is only relative and is plotted as a percentage of the maximum possible. Notice that the GaAsP LED provides the highest output at approximately 6600 angstroms and the output rapidly drops off on either side of this maximum point. This relatively narrow spectral output produces a red light as explained earlier.

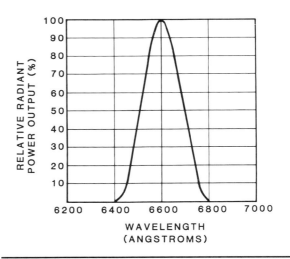

Figure 10-23 Spectral response of a typical GaAsP LED.

The performance of an LED can be expressed in terms of its efficiency or the number of output photons produced for each input electron. When its efficiency is expressed in this manner, it is found to be extremely low for all types of LED's. This low efficiency occurs for two basic reasons. First, much of the light energy produced never reaches the surface of the device where it can escape. It is simply absorbed by the semitransparent semiconductor material. The longer the path through which the light must travel, the greater will be the absorption. This is why the P-type region is made very thin. The thin P-type layer places the junction near the surface where light can escape. Unfortunately the emitted photons tend to move out in all directions from the junction. However, this situation is corrected by the gold contact or backing (refer to Figure 10-20) on the bottom of the chip. This backing reflects the light emitted toward the bottom of the chip back towards the P-type layer and helps to increase the efficiency of the device.

The low efficiency of an LED also results because much of the light is reflected back into the device. In order for the photons to escape, they must strike the surface within a specified critical angle as shown in Figure 10-24. For the GaAsP LED the critical angle is 17 degrees. This means that photons must strike the surface within a 17 degree angle (from the line drawn perpendicular to the surface) or be reflected back into the structure. The critical angle is true only when the photons are emitted into the air and under those conditions only 8 percent of the light produced can escape.

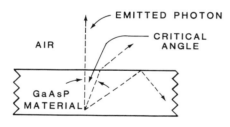

Figure 10-24 Relative paths of photons arriving at the surface of an LED chip.

When the LED chip is placed in a plastic (epoxy is sometimes used) package the critical angle is greatly increased and in some cases can be doubled. The package therefore increases the amount of light emitted in addition to magnifying or diffusing the light so that it can be easily seen.

As you may suspect, the amount of light produced by an LED is small when compared to an incandescent lamp. Most LED's generate a typical luminous intensity of only a few millicandelas, which is very low compared to even a miniature incandescent panel light which can produce many times that much light. However, LED's have several important advantages. First, they are extremely rugged. They also respond very quickly to changes in operating current and therefore can operate at extremely high speeds. They require very low operating voltages and are compatible with integrated circuits, transistors, and other solid-state devices. They are relatively inexpensive when compared to incandescent devices. Also, they may be designed to emit a specific light color or narrow frequency range when compared to the incandescent lamp, which emits a white light that contains a broad range of light frequencies.

The disadvantages associated with LED's (in addition to low light output) are similar to those which pertain to many types of solid-state components. They may be easily damaged by excessive voltage or current (beyond their maximum ratings) and their output radiant power is dependent on temperature.

LED Applications

In any application of an LED, the device is seldom used alone. The LED is usually connected in series with a resistor which limits the current flowing through the LED to the desired value. To operate the LED without this current limiting resistor would be risky since even a slight increase in operating voltage might cause an excessive amount of current to flow through the device. Some LED packages even contain built-in resistors (in chip form).

A schematic symbol that is commonly used to represent the LED is shown in Figure 10-25A and the correct way to bias an LED is shown in Figure 10-25B. The series resistor (R_s) must have a value which will limit the forward current (I_F) to the desired value based on the applied voltage (E) and the voltage drops across the LED which we assume to be equal to 1.6 volts. When determining the value of R_s allowance must also be made for the internal resistance of the LED (which has a typical value of 5 ohms). All of these factors are included in the following equation which can be used to determine the required value of R_s:

$$R_s = \frac{E - 1.6\,V}{I_F} - 5\,\Omega$$

If we assume that E is equal to 6 volts and that I_F must equal 50 milliamperes (0.05 amperes) to obtain the desired light intensity, R_s must be equal to:

$$R_S = \frac{6\,V - 1.6\,V}{0.05\,A} - 5\,\Omega$$

$$= 88\,\Omega - 5\,\Omega$$

$$= 83\,\Omega$$

The previous equation is useful for determining R_S as long as the required R_S value is equal to or greater than 40 ohms.

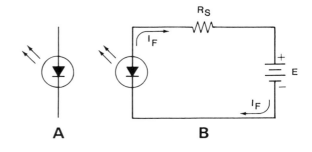

Figure 10-25 An LED symbol and a properly biased LED circuit.

Visible light producing LED's with their respective series resistors are often used as simple indicator lights to provide on and off indications. Individual LED's may even be arranged in specific patterns as shown in Figure 10-26. The LED's shown in this figure each illuminate one of seven segments arranged in a special pattern. The segments can be turned on or off to create the numbers 0 through 9 and certain letters. Such devices are referred to as 7 segment LED displays.

Figure 10-26 Typical segment LED numeric displays in a standard dual in line package. (Courtesy of Hewlett-Packard)

LED's that emit infrared light may be used in intrusion detector systems if the light is properly focused and controlled. The infrared light cannot be seen by the human eye and is very effective in this application.

7-Segment Displays

The 7-segment display is a group of LEDs arranged in a single chip. They may share either common cathode or common anode and be activated by applying either a ground or a positive pulse. They usually require an external driver (decoder chip) and external current limiting resistors.

When combined with an encoder/decoder chip they are capable of displaying 16 characters. These characters include the decimal numbers 0 through 9 and A, B, C, D, E, and F.

The 7-segment display comes in a variety of colors and by placing a plastic colored filter (thin colored plastic) over the chip a variety of colors and intensities can be achieved. The 7-segment display is used in both the Digital Techniques and Microprocessor courses.

Infrared LED's are also commonly used in conjunction with light-sensitive devices, such as photodiodes or phototransistors, to form what is called an optical coupler. A typical optical coupler, which utilizes an LED and a phototransistor, is shown in Figure 10-27. The LED and phototransistor chips are separated by a special type of light transmitting glass, and they are coupled only by the light beam produced by the LED. An electrical signal (varying current or voltage) applied to the LED'S terminals (through two of the pins on the mini-DIP package) will produce changes in the light beam which in turn varies the conductivity of the phototransistor. When properly biased, the phototransistor will convert the varying light energy back into an electrical signal. This type of arrangement allows a signal to pass from one circuit to another but provides a high degree of electrical isolation between the circuits. Also, the LED responds quickly to input signal changes, thus, making it possible to transmit high frequency AC signals through the optical coupler.

Figure 10-27 A typical optical coupler which contains an LED and a phototransistor. (*Courtesy of General Electric*)

Another application of the LED is shown in Figure 10-28. The device shown in this figure is called an optical limit switch. An LED is mounted in the right portion of the package and a photo-darlington circuit is mounted on the left side. They are separated by a narrow slot in the package. The device may be thought of as an optical coupler with an exposed light beam. When an object (such as a dime, as shown) is inserted into the slot, the light beam is broken and the photo-darlington circuit does not provide an output signal. This condition is identified by an LED indicator lamp which is mounted above the package. When the object is removed the light beam is restored and the indicator lamp turns on as shown in the right photo. Such a device is useful for sensing the presence of objects such as cards, tickets or tapes.

LED's are also used in conjunction with photodiodes or phototransistors to sense the presence of holes or perforations in paper tapes or cards. However, in this application a number of LED's as well as photodiodes or phototransistors are required and they are arranged in specific patterns to correspond with the holes in the tape or card.

Figure 10-28 An optical limit switch which senses the presence or absence of an object.
(Courtesy of Monsanto)

Desk-Top Experiment 5
Analysis of an LED Circuit

Introduction

This experiment will provide the experience necessary for you to analyze a basic LED circuit and maintain proper operation under a variety of operating conditions.

Objectives

1. Calculate the proper value of R_S under various conditions.

2. Draw a circuit that will function properly with a variety of supply voltages.

3. Analyze the operation of an LED circuit.

Procedure

Design an LED circuit that has a V_{DC} supply that varies from 0 to 15 volts. The LED is assumed to drop a constant 1.6 volts when forward biased and it has a maximum current rating of 20 mA.

1. Calculate the proper value for R_S when the supply voltage is set to 15 volts.

$$R_S = \underline{\hspace{2cm}}$$

2. To maintain the same LED brilliance when the input voltage is reduced to 10 volts, the value of R_S should be:

$$R_S = \underline{\hspace{2cm}}$$

3. To maintain the same LED brilliance when the input voltage is reduced to 5 volts, the value of R_S should be:

$$R_S = \underline{\hspace{2cm}}$$

4. Draw a circuit that could be used to satisfy the requirements listed in steps 1, 2, and 3.

5. With reference to the equipment and parts required to complete this course the logical choice for RS is:

Discussion

The supply voltage minus the LED voltage drop divided by $I_{F(max)}$ is used to calculate the value of R_S.

$$R_S = \frac{V_{DC} - 1.6\,V}{20\,mA}$$

The circuit that you drew in step 4 should resemble the circuit shown in Figure 10-29. This circuit provides a variable DC supply and an adjustable R_S.

When the supply voltage decreased, the value of RS should have been decreased to maintain a constant 1.6 volt supply to the LED. This would have maintained the current flow through the LED at approximately 20 mA.

Since R_S has 3 distinct values that range from more than 150 ohms to less than 800 ohms the logical choice for R_S is the 1 k ohm potentiometer located on the trainer.

Procedure (Cont.)

In reference to Figure 10-29.

6. R$_S$ is the resistance between terminals?

    _____ and _____

7. To increase the value of R$_S$ terminal _____ should be moved toward terminal _____.

8. What is the required voltage drop between terminals 1 and 2.

    _____

9. What is the required voltage drop between terminals 2 and 3.

    _____

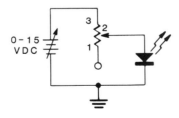

Figure 10-29 Circuit for Desk-Top Experiment 5.

Discussion

R$_S$ is the resistance that is in series with the LED and in this case, it must restrict current flow through the LED to slightly less than 20 mA. The resistance between terminals 2 and 3 is R$_S$.

To increase the value of R$_S$ requires that terminal 2 be moved toward terminal 1. This will result in a decrease in brilliance unless the supply voltage is increased.

The required voltage drop between terminals 1 and 2 is 1.6 volts. Note, that terminals 1 and 2 are in parallel with the LED.

The voltage drop between terminals 2 and 3 is the supply voltage minus the voltage drop across the LED. Note, also that the current flowing through R$_S$ is higher than the current flowing through the LED. This is caused by the parallel path from terminal 1. Therefore, the value of R$_S$ could be smaller than calculated.

Procedure (Cont.)

10. Center the potentiometer and reverse the LED in the circuit. What is the LED current?

    I$_{LED}$ =

11. The voltage drop across the potentiometer will be:

_____

12. The voltage drop from terminal 2 to ground is:

_____

Discussion

When the LED is reversed in the circuit it will not conduct (reverse biased). The LED will appear to be an open circuit and there will be no current flow. Since current is not flowing the LED will not emit light.

The voltage drop across the potentiometer will be V_{DC}. There is a complete path for current flow even with the LED reverse biased.

The voltage drop from terminal 2 to ground will be V_{DC} divided by 2. Remember that in step 10 you centered the potentiometer.

$$\frac{V_{DC}}{2}$$

Summary

The circuit requirements are a variable DC supply and an adjustable R_S that maintains the current through the LED to less than 20 mA. The circuit shown in Figure 10-29 could be used for that purpose.

R_S could have been a fixed resistor but you would have to physically remove and reinsert the proper value of R_S for each value of supply voltage selected. This also presents another problem. After each calculation you will have to select the next higher standard value resistor.

In Figure 10-29 it is easier to adjust the potentiometer to maximum resistance, set the supply voltage to the desired value, and then adjust the potentiometer using the LED's brilliance as your indicator. This provides a simple circuit (fewer parts) that is variable (both V_{DC} and R_S) and relatively safe to operate (LED overload).

Liquid Crystal Displays

Liquid-crystal displays are usually packaged as numeric or alphanumeric displays similar to those often used to package LED's. However, the liquid crystal display (LCD) differs from an LED display in one important respect, LED's generate light while LCD's control light. Consequently, the appearance of an LCD is quite different from that of an LED. LCD characters are usually dark against a light background. A significant advantage of LCD displays is that they have very low operating voltage and current requirements. As a result, these displays have become very popular in digital wrist watches, test equipment, calculators, and other battery powered devices. Figure 10-30 shows a portable multimeter that uses an LCD readout.

The term liquid crystal may seem to be contradictory because a crystalline structure is usually associated with solids. However, there is an unusual group of fluids that has a molecular structure like that of crystalline solids. The fluids in this group that are used in optoelectronic displays are nematic liquid crystals. Nematic liquid crystals have long molecules that are ordinarily aligned in a specific pattern. As shown in Figure 10-31A, light can readily pass through this molecular structure. Therefore, the material is transparent. However, as indicated in Figure 10-31B, if an electrical or magnetic field is applied to all or part of the liquid crystal, the molecular arrangement in that part of the crystal is subjected to the field changes. This molecular rearrangement causes scattered reflection of light with the result that the material takes on an opaque, frosty appearance.

Figure 10-30 Typical hand-held multimeter, using an LCD readout.

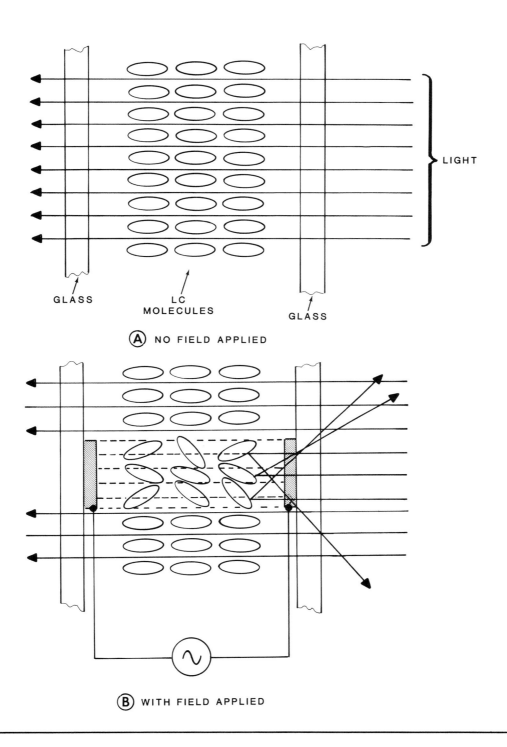

Figure 10-31 Dynamic scattering effect.

Figure 10-32 shows the construction of a typical liquid-crystal display. As shown, the liquid crystal material is sealed between two layers of glass. Transparent electrodes have been deposited on each layer of glass. In addition, the molecules in the nematic liquid crystal are aligned in a uniform lattice arrangement that is perpendicular to the glass, allowing light to pass through the display.

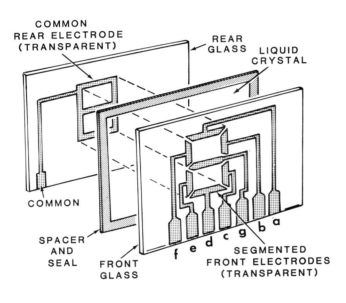

Figure 10-32 Exploded view of LCD showing construction details.

Application of a voltage to one or more front elctrodes and the rear electrode produces an electrostatic field, forcing the liquid crystal molecules within the electrostatic field to rearrange themselves into a random alignment pattern. As a result, light striking this part of the display is scattered, causing it to appear opaque. With the addition of a dark, light absorbing background to the display, frosty white characters will stand out clearly.

LCD's have the disadvantage that they must be illuminated by an external light source. Consequently, they cannot be read in the dark or in low light levels. In some digital wrist watches, and other similar devices, this disadvantage is partially overcome by installing a small light source in or near the display. This allows the user to switch the supply on and off as required to read the display.

The LCDs are relatively easy to read in normal room light and have the advantage over many displays that the brighter the light the easier they are to read. Some LCDs also have the advantage of being read from either side while most visual displays can only be read from one side. This of course would depend on its mounting.

29. Light-emitting devices convert electrical energy into _____ energy.

20. Incandescent lamps have a very long life expectancy.

    A. True
    B. False

31. The light-emitting diode must be forward biased to operate properly.

    A. True
    B. False

32. A properly biased LED emits light because of the recombination of _____ and _____ near its PN junction.

33. LED's made with gallium arsenide emit light in the _____ region.

34. Gallium phosphide LED's produce a visible _____ light.

35. The LED's top electrical contact may have long protrusions or fingers as shown in Figure 10-20B so that the current through its junctions will be evenly _____.

36. The LED package usually has a dome shaped top which serves as a _____.

37. The radiant power produced by an LED increases linearly with its input _____.

38. The low efficiency of an LED results because much of the light generated within the device is either absorbed or _____ back into its structure.

39. In order for the photons to escape from the surface of an LED they must strike the surface within a specified _____ _____.

40. A _____ is normally placed in series with an LED to limit its current to the desired value.

41. LED's that are arranged into patterns so that they will create numbers are refered to as LED _____.

42. An optical coupler containing an LED and a photodiode will transmit information over a light beam but at the same time provide a high degree of electrical _____.

43. LCD's are easier to read in bright sun light than the 7-segment displays.

    A. True.
    B. False.

Optoelectronic devices are now widely used to perform various functions in electronic equipment. They may be divided into two general categories: light-sensitive devices and light-emitting devices. The photodiode has the fastest response time.

The light-sensitive category includes devices such as photoconductive cells, photovoltaic cells, photodiodes, and phototransistors. These devices respond to changes in light intensity by either generating an output voltage or by changing their internal resistance.

The light-emitting category includes components such as incandescent and neon lamps, but these older components are no longer widely used. They are being replaced by a solid-state device known as a light-emitting diode or LED.

LED's can be designed to emit light over narrowly combined regions of the light spectrum when subjected to relatively low currents and voltages.

Infrared emitting diodes are special purpose LED's that emit energy in the infrared region of the spectrum. They have characteristics which make them highly compatible with other types of solid-state devices such as integrated circuits and transistors. They also provide many of the same benefits that are realized with most types of solid-state devices.

There are many applications for optoelectronic devices. These components may be used individually, but they can be combined to perform a variety of functions. The LED is often combined with a photodiode or phototransistor to provide optical coupling between circuits or to detect the presence of an object. Individual LED's are used as indicators and combinations of LED's are used to display numbers. The number of possible applications are endless. The liquid-crystal display is another numeric display device that has applications similar to those for LED displays. The LCD is very popular in applications where low power consumption is an important consideration. The true potential of these important components are just now being realized and continued growth in the optoelectronics field is almost certain.

The LCD uses low voltage and low currents and can be easily read in bright light. In fact, the brighter the light the easier the display is to read. It also has the advantage of being visible from either side of the display. However, it requires a controlled beam to form the display.

APPENDIX A
DESK-TOP EXPERIMENT 6

Desk-Top Experiment 6
Identifying Schematic Symbols for Semiconductor Components

Introduction

Now that you have completed all of the units in this text book, it is time to demonstrate how much of the material you retained. In this experiment you will identify a variety of schematic solid state semiconductor symbols, explain their applications, and make comparisons between selected symbols.

Objectives

1. Identify semiconductor devices by their schematic symbols.

2. Identify the schematic symbols that are light dependent.

3. Draw the amplifier symbols and their inputs and respected outputs.

4. Draw the symbols that are digital logic gates and explain the application where each gate would be used. You may explain in words or by using the symbols with proper inputs and outputs.

5. Explain the advantages and/or applications for the selected semiconductor devices.

Procedure

1. Identify the schematic symbols shown in Figure 10-33. Be as specific as possible. Figure 10-33 is located on the last page of this experiment.

For example

A. is a PNP bipolar transistor
B.
C.
D.
E.
F.
G.
H.
I.
J.
K.
L.
M.
N.
O.
P.
Q.
R.
S.
T.
U.
V.
W.
X.
Y.
Z.

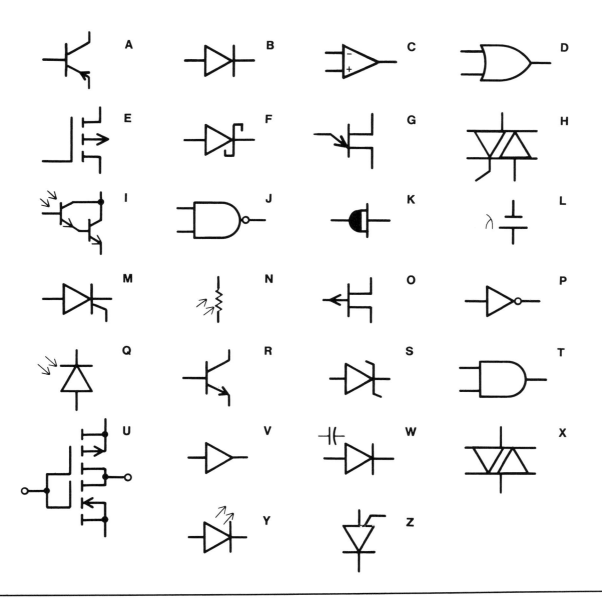

Figure 10-33 Schematic symbols for Desk-Top Experiment 6.

Discussion

There are 26 schematic symbols shown in Figure 10-33. They are all from the units in the Semiconductor Devices course and should be easily recognized by every student completing the course.

Procedure (Cont.)

2. List the schematic symbols that represent components that are light dependent.

3. List the amplifier symbols and explain the application for each (why one is preferred over the others).

4. List the digital logic gate symbols and describe when each would be used (what function does one gate provide that the other gates can't).

Discussion

There are 5 symbols that represent light sensitive components, but only 4 of the components are light dependent.

There are 4 symbols that actually represent amplifiers, 3 are capable of amplifing voltage and 1 is considered to be a current amplifier.

There are 4 symbols that represent digital logic gates. They are the AND, OR, NOT, and NAND gates.

Procedure (Cont.)

5. Draw the schematic symbols complete with input and output waveforms for the symbols indicated:

    B.

    C.

    P.

    T.

6. Which schematic symbol represents a component that is normally considered to be a rectifier?

   What is a rectifier?

7. Which schematic symbol represents a voltage regulator?

   What type bias is required for proper operation?

8. Which schematic symbol represents a component that is normally operated in its negative resistance region.

   What does the term negative resistance mean?

9. Which of the schematic symbols use the gate to turn on the component but not to turn the component off.

   What polarity pulse is required to turn the device on?

10. Which of the schematic symbols represents devices that use reverse bias to control their interelectrode capacitance?

    Increasing the reverse bias causes the internal capacitance to:

    _____

11. Explain the advantage or application of the following components:

    H.

    G.

    M.

    Z.

    What do the symbols have in common?

12. What is the primary advantage of the component represented by symbol F?

13. Draw the I-V curves for the components represented by symbols B and S and label the normal conduction points.

14. Which component in Figure 10-33 contains a complementary pair?

Summary

This experiment is a test of your understanding of the material presented in the Semiconductor Devices course. If you can identify the component symbols, explain their operational function, and differentiate between why components in the same family are preferred for a particular application, you will have gained a great deal of information concerning solid state devices and their applications. This experiment was designed as a course review and should be completed prior to the final examination.

If you had any problems with this experiment you should reread the particular unit that contains the information causing your confusion.

APPENDIX B
DATA SHEETS

INTRODUCTION

This appendix is a collection of data sheets for some of the components discussed and used in the **Semiconductor Devices Course EB-6103A.** They were furnished courtesy of **MOTOROLA, HUGHES, RCA, and INTERSEL.**

CONTENTS

MOTOROLA

SEMICONDUCTORS

P.O. BOX 20912 • PHOENIX, ARIZONA 85036

Designers▲Data Sheet

1N4001
thru
1N4007

"SURMETIC"▲ RECTIFIERS

. . . subminiature size, axial lead mounted rectifiers for general-purpose low-power applications.

Designers Data for "Worst Case" Conditions

The Designers▲ Data Sheets permit the design of most circuits entirely from the information presented. Limit curves — representing boundaries on device characteristics — are given to facilitate "worst case" design.

LEAD MOUNTED
SILICON RECTIFIERS

50-1000 VOLTS
DIFFUSED JUNCTION

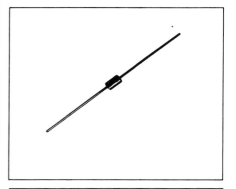

*MAXIMUM RATINGS

Rating	Symbol	1N4001	1N4002	1N4003	1N4004	1N4005	1N4006	1N4007	Unit
Peak Repetitive Reverse Voltage Working Peak Reverse Voltage DC Blocking Voltage	V_{RRM} V_{RWM} V_R	50	100	200	400	600	800	1000	Volts
Non-Repetitive Peak Reverse Voltage (halfwave, single phase, 60 Hz)	V_{RSM}	60	120	240	480	720	1000	1200	Volts
RMS Reverse Voltage	$V_{R(RMS)}$	35	70	140	280	420	560	700	Volts
Average Rectified Forward Current (single phase, resistive load, 60 Hz, see Figure 8, T_A = 75°C)	I_O	1.0							Amp
Non-Repetitive Peak Surge Current (surge applied at rated load conditions, see Figure 2)	I_{FSM}	30 (for 1 cycle)							Amp
Operating and Storage Junction Temperature Range	T_J, T_{stg}	−65 to +175							°C

*ELECTRICAL CHARACTERISTICS

Characteristic and Conditions	Symbol	Typ	Max	Unit
Maximum Instantaneous Forward Voltage Drop (i_F = 1.0 Amp, T_J = 25°C) Figure 1	v_F	0.93	1.1	Volts
Maximum Full-Cycle Average Forward Voltage Drop (I_O = 1.0 Amp, T_L = 75°C, 1 inch leads)	$V_{F(AV)}$	—	0.8	Volts
Maximum Reverse Current (rated dc voltage) T_J = 25°C T_J = 100°C	I_R	0.05 1.0	10 50	μA
Maximum Full-Cycle Average Reverse Current (I_O = 1.0 Amp, T_L = 75°C, 1 inch leads	$I_{R(AV)}$	—	30	μA

*Indicates JEDEC Registered Data.

MECHANICAL CHARACTERISTICS

CASE: Transfer Molded Plastic
MAXIMUM LEAD TEMPERATURE FOR SOLDERING PURPOSES: 350°C, 3/8'' from case for 10 seconds at 5 lbs. tension
FINISH: All external surfaces are corrosion-resistant, leads are readily solderable
POLARITY: Cathode indicated by color band
WEIGHT: 0.40 Grams (approximately)

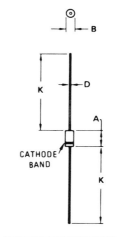

CATHODE
BAND

DIM	MILLIMETERS		INCHES	
	MIN	MAX	MIN	MAX
A	5.97	6.60	0.235	0.260
B	2.79	3.05	0.110	0.120
D	0.76	0.86	0.030	0.034
K	27.94	—	1.100	—

CASE 59-04
Does Not Conform to DO-41 Outline.

▲Trademark of Motorola Inc.

© MOTOROLA INC., 1982

DS 6015 R3

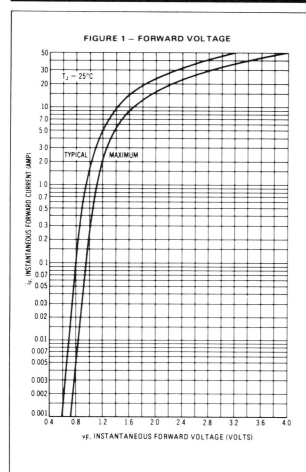

FIGURE 1 — FORWARD VOLTAGE

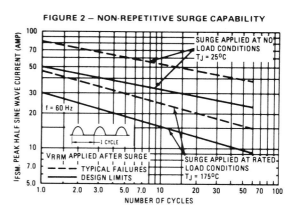

FIGURE 2 — NON-REPETITIVE SURGE CAPABILITY

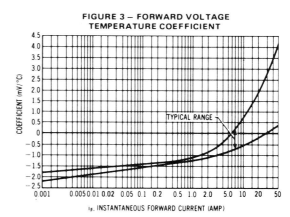

FIGURE 3 — FORWARD VOLTAGE TEMPERATURE COEFFICIENT

FIGURE 4 — TYPICAL TRANSIENT THERMAL RESISTANCE

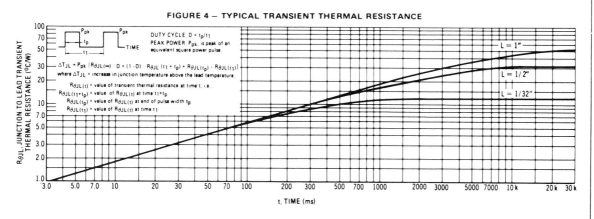

The temperature of the lead should be measured using a thermocouple placed on the lead as close as possible to the tie point. The thermal mass connected to the tie point is normally large enough so that it will not significantly respond to heat surges generated in the diode as a result of pulsed operation once steady-state conditions are achieved. Using the measured value of T_L, the junction temperature may be determined by:

$$T_J = T_L + \triangle T_{JL}.$$

 MOTOROLA *Semiconductor Products Inc.*

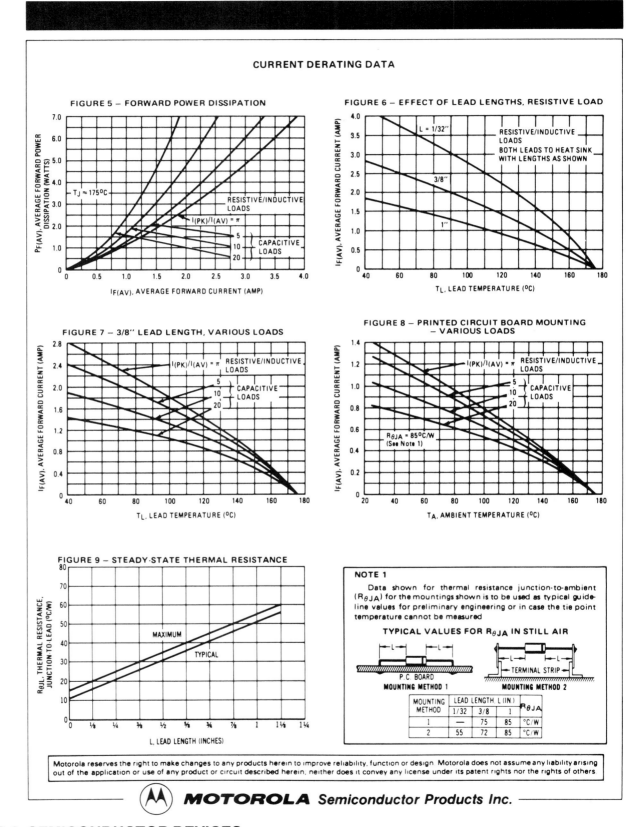

CURRENT DERATING DATA

FIGURE 5 — FORWARD POWER DISSIPATION

FIGURE 6 — EFFECT OF LEAD LENGTHS, RESISTIVE LOAD

FIGURE 7 — 3/8" LEAD LENGTH, VARIOUS LOADS

FIGURE 8 — PRINTED CIRCUIT BOARD MOUNTING — VARIOUS LOADS

FIGURE 9 — STEADY-STATE THERMAL RESISTANCE

NOTE 1

Data shown for thermal resistance junction-to-ambient ($R_{\theta JA}$) for the mountings shown is to be used as typical guideline values for preliminary engineering or in case the tie point temperature cannot be measured

TYPICAL VALUES FOR $R_{\theta JA}$ IN STILL AIR

MOUNTING METHOD 1

MOUNTING METHOD 2

MOUNTING METHOD	LEAD LENGTH, L (IN.)			$R_{\theta JA}$
	1/32	3/8	1	
1	—	75	85	°C/W
2	55	72	85	°C/W

Motorola reserves the right to make changes to any products herein to improve reliability, function or design. Motorola does not assume any liability arising out of the application or use of any product or circuit described herein, neither does it convey any license under its patent rights nor the rights of others.

MOTOROLA *Semiconductor Products Inc.*

B-6 SEMICONDUCTOR DEVICES

TYPICAL DYNAMIC CHARACTERISTICS

FIGURE 10 — FORWARD RECOVERY TIME

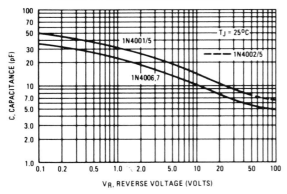

FIGURE 11 — REVERSE RECOVERY TIME

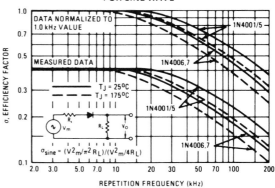

FIGURE 12 — JUNCTION CAPACITANCE

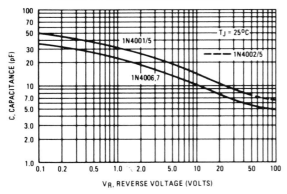

FIGURE 13 — RECTIFICATION WAVEFORM EFFICIENCY FOR SINE WAVE

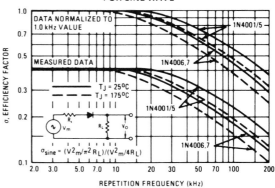

FIGURE 14 — RECTIFICATION WAVEFORM EFFICIENCY FOR SQUARE WAVE

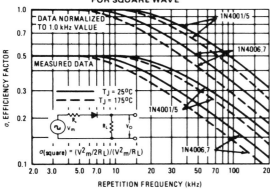

RECTIFIER EFFICIENCY NOTE

The rectification efficiency factor σ shown in Figures 13 and 14 was calculated using the formula:

$$\sigma = \frac{P_{dc}}{P_{rms}} = \frac{\dfrac{V^2_{O(dc)}}{R_L}}{\dfrac{V^2_{O(rms)}}{R_L}} \bullet 100\% = \frac{V^2_{O(dc)}}{V^2_{O(ac)} + V^2_{O(dc)}} \bullet 100\% \quad (1)$$

For a sine wave input $V_m\sin(\omega t)$ to the diode, assumed lossless, the maximum theoretical efficiency factor becomes 40%, for a square wave input of amplitude V_m, the efficiency factor becomes 50%. (A full wave circuit has twice these efficiencies).

As the frequency of the input signal is increased, the reverse recovery time of the diode (Figure 11) becomes significant, resulting in an increasing ac voltage component across R_L which is opposite in polarity to the forward current thereby reducing the value of the efficiency factor σ, as shown in Figures 13 and 14.

It should be emphasized that Figures 13 and 14 show waveform efficiency only; they do not account for diode losses. Data was obtained by measuring the ac component of V_O with a true rms voltmeter and the dc component with a dc voltmeter. The data was used in Equation 1 to obtain points for the Figures.

 MOTOROLA *Semiconductor Products Inc.*

BOX 20912 • PHOENIX, ARIZONA 85036 • A SUBSIDIARY OF MOTOROLA INC.

 **MOTOROLA**

Designers Data Sheet

500 MILLIWATT HERMETICALLY SEALED GLASS SILICON ZENER DIODES

- Complete Voltage Range — 2.4 to 110 Volts**
- DO-35 Package — Smaller than Conventional DO-7 Package
- Double Slug Type Construction
- Metallurgically Bonded Construction
- Nitride Passivated Die

Designer's Data for "Worst Case" Conditions

The Designer's Data sheets permit the design of most circuits entirely from the information presented. Limit curves — representing boundaries on device characteristics — are given to facilitate "worst case" design.

1N5221 thru 1N5272

GLASS ZENER DIODES

500 MILLIWATTS
2.4-110 VOLTS

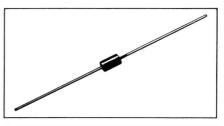

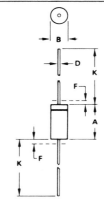

*MAXIMUM RATINGS

Rating	Symbol	Value	Unit
DC Power Dissipation @ $T_L \leq 75^\circ C$	P_D		
Lead Length = 3/8"		500	mW
Derate above $T_L = 75^\circ C$		4.0	mW/$^\circ$C
Operating and Storage Junction Temperature Range	T_J, T_{stg}	−65 to + 200	$^\circ$C

*Indicates JEDEC Registered Data
**See 1N5273 thru 1N5281 for devices > 110 volts.

MECHANICAL CHARACTERISTICS

CASE: Double slug type, hermetically sealed glass

MAXIMUM LEAD TEMPERATURE FOR SOLDERING PURPOSES: 230°C, 1/16" from case for 10 seconds

FINISH: All external surfaces are corrosion resistant with readily solderable leads

POLARITY: Cathode indicated by color band. When operated in zener mode, cathode will be positive with respect to anode

MOUNTING POSITION: Any

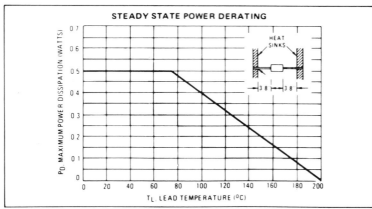

STEADY STATE POWER DERATING

Designer's is a trademark of Motorola Inc.

NOTES:
1. PACKAGE CONTOUR OPTIONAL WITHIN A AND B. HEAT SLUGS, IF ANY, SHALL BE INCLUDED WITHIN THIS CYLINDER, BUT NOT SUBJECT TO THE MINIMUM LIMIT OF B.
2. LEAD DIAMETER NOT CONTROLLED IN ZONE F TO ALLOW FOR FLASH, LEAD FINISH BUILDUP AND MINOR IRREGULARITIES OTHER THAN HEAT SLUGS.
3. POLARITY DENOTED BY CATHODE BAND.
4. DIMENSIONING AND TOLERANCING PER ANSI Y14.5, 1973.

DIM	MILLIMETERS		INCHES	
	MIN	MAX	MIN	MAX
A	3.05	5.08	0.120	0.200
B	1.52	2.29	0.060	0.090
D	0.46	0.56	0.018	0.022
F	–	1.27	–	0.050
K	25.40	38.10	1.000	1.500

All JEDEC dimensions and notes apply.

CASE 299-02
DO-204AH
(DO-35)

© MOTOROLA INC. 1982   DS 7051R1

ELECTRICAL CHARACTERISTICS

(T_A = 25°C unless otherwise noted. Based on dc measurements at thermal equilibrium; lead length = 3/8"; thermal resistance of heat sink = 30°C/W) V_F = 1.1 max @ I_F = 200 mA for all types.

JEDEC Type No. (Note 1)	Nominal Zener Voltage V_Z @ I_{ZT} Volts (Note 2)	Test Current I_{ZT} mA	Max Zener Impedance A and B Suffix only		Max Reverse Leakage Current				Max Zener Voltage Temperature Coeff. (A and B Suffix only) θ_{VZ} (%/°C) (Note 3)
					A and B Suffix only			Non-Suffix	
			Z_{ZT} @ I_{ZT} Ohms	Z_{ZK} @ I_{ZK} = 0.25 mA Ohms	I_R µA	@ V_R Volts A	B	I_R @ V_R Used for Suffix A µA	
1N5221	2.4	20	30	1200	100	0.95	1.0	200	−0.085
1N5222	2.5	20	30	1250	100	0.95	1.0	200	−0.085
1N5223	2.7	20	30	1300	75	0.95	1.0	150	−0.080
1N5224	2.8	20	30	1400	75	0.95	1.0	150	−0.080
1N5225	3.0	20	29	1600	50	0.95	1.0	100	−0.075
1N5226	3.3	20	28	1600	25	0.95	1.0	100	−0.070
1N5227	3.6	20	24	1700	15	0.95	1.0	100	−0.065
1N5228	3.9	20	23	1900	10	0.95	1.0	75	−0.060
1N5229	4.3	20	22	2000	5.0	0.95	1.0	50	−0.055
1N5230	4.7	20	19	1900	5.0	1.9	2.0	50	±0.030
1N5231	5.1	20	17	1600	5.0	1.9	2.0	50	−0.030
1N5232	5.6	20	11	1600	5.0	2.9	3.0	50	+0.038
1N5233	6.0	20	7.0	1600	5.0	3.3	3.5	50	+0.038
1N5234	6.2	20	7.0	1000	5.0	3.8	4.0	50	+0.045
1N5235	6.8	20	5.0	750	3.0	4.8	5.0	30	+0.050
1N5236	7.5	20	6.0	500	3.0	5.7	6.0	30	+0.058
1N5237	8.2	20	8.0	500	3.0	6.2	6.5	30	+0.062
1N5238	8.7	20	8.0	600	3.0	6.2	6.5	30	+0.065
1N5239	9.1	20	10	600	3.0	6.7	7.0	30	+0.068
1N5240	10	20	17	600	3.0	7.6	8.0	30	+0.075
1N5241	11	20	22	600	2.0	8.0	8.4	30	+0.076
1N5242	12	20	30	600	1.0	8.7	9.1	10	+0.077
1N5243	13	9.5	13	600	0.5	9.4	9.9	10	+0.079
1N5244	14	9.0	15	600	0.1	9.5	10	10	+0.082
1N5245	15	8.5	16	600	0.1	10.5	11	10	+0.082
1N5246	16	7.8	17	600	0.1	11.4	12	10	+0.083
1N5247	17	7.4	19	600	0.1	12.4	13	10	+0.084
1N5248	18	7.0	21	600	0.1	13.3	14	10	+0.085
1N5249	19	6.6	23	600	0.1	13.3	14	10	+0.086
1N5250	20	6.2	25	600	0.1	14.3	15	10	+0.086
1N5251	22	5.6	29	600	0.1	16.2	17	10	+0.087
1N5252	24	5.2	33	600	0.1	17.1	18	10	+0.088
1N5253	25	5.0	35	600	0.1	18.1	19	10	+0.089
1N5254	27	4.6	41	600	0.1	20	21	10	+0.090
1N5255	28	4.5	44	600	0.1	20	21	10	+0.091
1N5256	30	4.2	49	600	0.1	22	23	10	+0.091
1N5257	33	3.8	58	700	0.1	24	25	10	+0.092
1N5258	36	3.4	70	700	0.1	26	27	10	+0.093
1N5259	39	3.2	80	800	0.1	29	30	10	+0.094
1N5260	43	3.0	93	900	0.1	31	33	10	+0.095
1N5261	47	2.7	105	1000	0.1	34	36	10	+0.095
1N5262	51	2.5	125	1100	0.1	37	39	10	+0.096
1N5263	56	2.2	150	1300	0.1	41	43	10	+0.096
1N5264	60	2.1	170	1400	0.1	44	46	10	+0.097
1N5265	62	2.0	185	1400	0.1	45	47	10	+0.097
1N5266	68	1.8	230	1600	0.1	49	52	10	+0.097
1N5267	75	1.7	270	1700	0.1	53	56	10	+0.098
1N5268	82	1.5	330	2000	0.1	59	62	10	+0.098
1N5269	87	1.4	370	2200	0.1	65	68	10	+0.099
1N5270	91	1.4	400	2300	0.1	66	69	10	+0.099
1N5271	100	1.3	500	2600	0.1	72	76	10	+0.110
1N5272	110	1.1	750	3000	0.1	80	84	10	+0.110

NOTE 1. Tolerance — The JEDEC type numbers shown indicate a tolerance of ±10% with guaranteed limits on only V_Z, I_R and V_F as shown in the electrical characteristics table. Units with guaranteed limits on all six parameters are indicated by suffix "A" for ±10% tolerance and suffix "B" for ±5.0% units.

†For more information on special selections contact your nearest Motorola representative.

NOTE 2. Special Selections† Available Include:

1. Nominal zener voltages between those shown.

2. Two or more units for series connection with specified tolerance on total voltage. Series matched sets make zener voltages in excess of 200 volts possible as well as providing lower temperature coefficients, lower dynamic impedance and greater power handling ability.

3. Nominal voltages at non-standard test currents.

 MOTOROLA *Semiconductor Products Inc.*

NOTE 3. Temperature Coefficient (θ_{VZ}) — Test conditions for temperature coefficient are as follows:

   a. I_{ZT} = 7.5 mA, T_1 = 25°C,
      T_2 = 125°C (1N5221A,B through 1N5242A,B).
   b. I_{ZT} = Rated I_{ZT}, T_1 = 25°C,
      T_2 = 125°C (1N5243A,B through 1N5272A,B).

Device to be temperature stabilized with current applied prior to reading breakdown voltage at the specified ambient temperature.

NOTE 4. Zener Voltage (V_Z) Measurement — Nominal zener voltage is measured with the device junction in thermal equilibrium at the lead temperature of 30°C ± 1°C and 3/8″ lead length.

NOTE 5. Zener Impedance (Z_Z) Derivation — Z_{ZT} and Z_{ZK} are measured by dividing the ac voltage drop across the device by the ac current applied. The specified limits are for $I_Z(ac)$ = $I_Z(dc)$ with the ac frequency = 60 Hz.

APPLICATION NOTE

Since the actual voltage available from a given zener diode is temperature dependent, it is necessary to determine junction temperature under any set of operating conditions in order to calculate its value. The following procedure is recommended:

Lead Temperature, T_L, should be determined from:

$$T_L = \theta_{LA} P_D + T_A.$$

θ_{LA} is the lead-to-ambient thermal resistance (°C/W) and P_D is the power dissipation. The value for θ_{LA} will vary and depends on the device mounting method. θ_{LA} is generally 30 to 40°C/W for the various clips and tie points in common use and for printed circuit board wiring.

The temperature of the lead can also be measured using a thermocouple placed on the lead as close as possible to the tie point. The thermal mass connected to the tie point is normally large enough so that it will not significantly respond to heat surges generated in the diode as a result of pulsed operation once steady-state conditions are achieved. Using the measured value of T_L, the junction temperature may be determined by:

$$T_J = T_L + \Delta T_{JL}.$$

ΔT_{JL} is the increase in junction temperature above the lead temperature and may be found from Figure 1 for dc power:

$$\Delta T_{JL} = \theta_{JL} P_D.$$

For worst-case design, using expected limits of I_Z, limits of P_D and the extremes of $T_J(\Delta T_J)$ may be estimated. Changes in voltage, V_Z, can then be found from:

$$\Delta V = \theta_{VZ} \Delta T_J.$$

θ_{VZ}, the zener voltage temperature coefficient, is found from Figures 3 and 4.

Under high power-pulse operation, the zener voltage will vary with time and may also be affected significantly by the zener resistance. For best regulation, keep current excursions as low as possible.

Surge limitations are given in Figure 6. They are lower than would be expected by considering only junc-

tion temperature, as current crowding effects cause temperatures to be extremely high in small spots, resulting in device degradation should the limits of Figure 6 be exceeded.

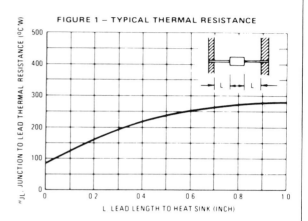

FIGURE 1 — TYPICAL THERMAL RESISTANCE

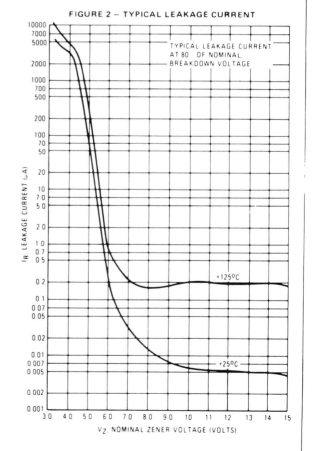

FIGURE 2 — TYPICAL LEAKAGE CURRENT

 MOTOROLA *Semiconductor Products Inc.*

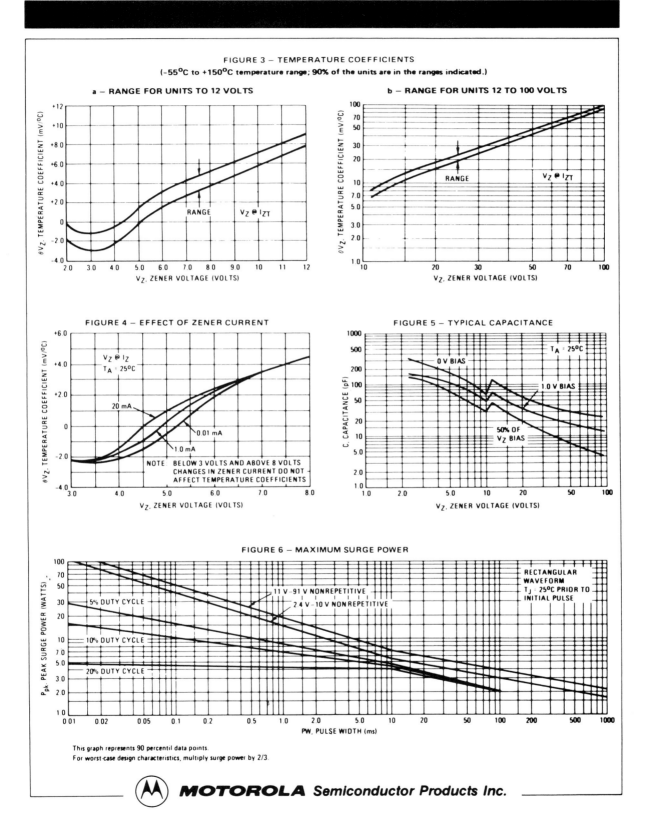

FIGURE 3 — TEMPERATURE COEFFICIENTS

(-55°C to +150°C temperature range; 90% of the units are in the ranges indicated.)

a — RANGE FOR UNITS TO 12 VOLTS

b — RANGE FOR UNITS 12 TO 100 VOLTS

FIGURE 4 — EFFECT OF ZENER CURRENT

FIGURE 5 — TYPICAL CAPACITANCE

FIGURE 6 — MAXIMUM SURGE POWER

This graph represents 90 percentil data points.

For worst-case design characteristics, multiply surge power by 2/3.

MOTOROLA *Semiconductor Products Inc.*

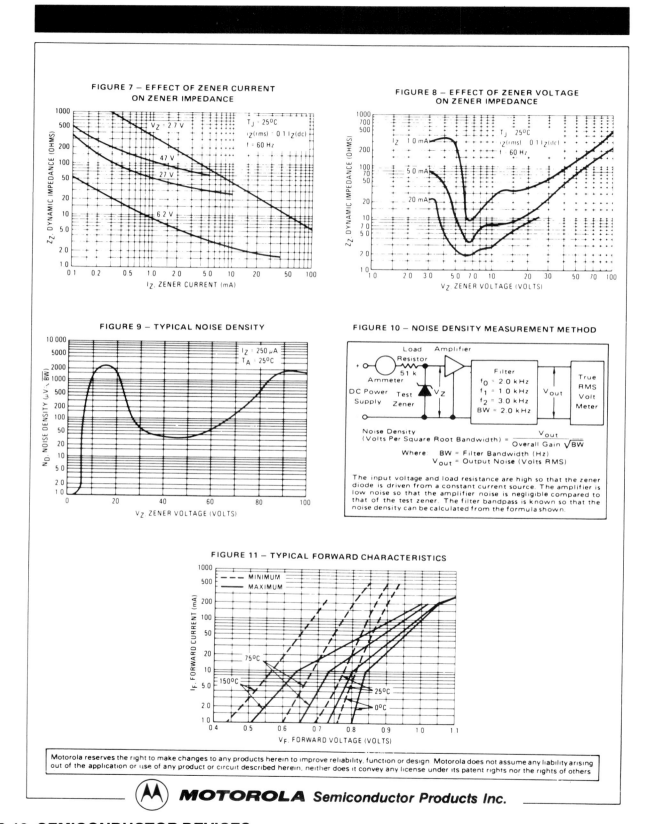

FIGURE 7 — EFFECT OF ZENER CURRENT
ON ZENER IMPEDANCE

FIGURE 8 — EFFECT OF ZENER VOLTAGE
ON ZENER IMPEDANCE

FIGURE 9 — TYPICAL NOISE DENSITY

FIGURE 10 — NOISE DENSITY MEASUREMENT METHOD

Noise Density
(Volts Per Square Root Bandwidth) = $\dfrac{V_{out}}{Overall\ Gain\ \sqrt{BW}}$

Where: BW = Filter Bandwidth (Hz)
V_{out} = Output Noise (Volts RMS)

The input voltage and load resistance are high so that the zener diode is driven from a constant current source. The amplifier is low noise so that the amplifier noise is negligible compared to that of the test zener. The filter bandpass is known so that the noise density can be calculated from the formula shown.

FIGURE 11 — TYPICAL FORWARD CHARACTERISTICS

Motorola reserves the right to make changes to any products herein to improve reliability, function or design. Motorola does not assume any liability arising out of the application or use of any product or circuit described herein, neither does it convey any license under its patent rights nor the rights of others.

Ⓜ **MOTOROLA** *Semiconductor Products Inc.*

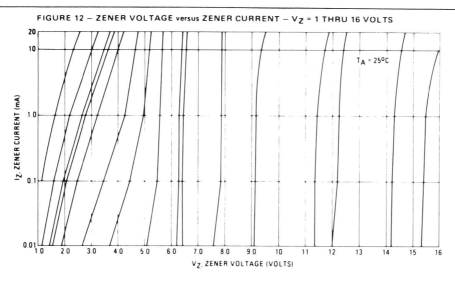

FIGURE 12 — ZENER VOLTAGE versus ZENER CURRENT — V_Z = 1 THRU 16 VOLTS

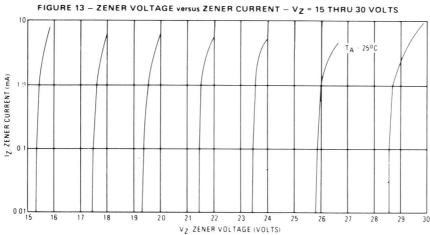

FIGURE 13 — ZENER VOLTAGE versus ZENER CURRENT — V_Z = 15 THRU 30 VOLTS

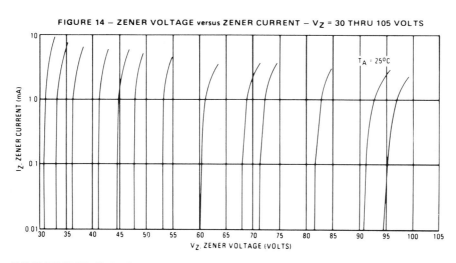

FIGURE 14 — ZENER VOLTAGE versus ZENER CURRENT — V_Z = 30 THRU 105 VOLTS

 MOTOROLA *Semiconductor Products Inc.*

BOX 20912 ● PHOENIX, ARIZONA 85036 ● A SUBSIDIARY OF MOTOROLA INC.

12096-2 PRINTED IN USA 7-82 IMPERIAL LITHO C07133 10,000   D67051-81

MOTOROLA

SEMICONDUCTORS

P.O. BOX 20912 • PHOENIX, ARIZONA 85036

LM139, A
LM239, A   LM2901
LM339, A   MC3302

QUAD SINGLE-SUPPLY COMPARATORS

These comparators are designed for use in level detection, low-level sensing and memory applications in Consumer Automotive and Industrial electronic applications.

- Single of Split Supply Operation
- Low Input Bias Current — 25 nA (Typ)
- Low Input Offset Current — ±5.0 nA (Typ)
- Low Input Offset Voltage — ±1.0 mV (Typ LM139A Series)
- Input Common-Mode Voltage Range to Gnd
- Low Output Saturation Voltage — 130 mV (Typ) @ 4.0 mA
- TTL and CMOS Compatible

QUAD COMPARATORS

SILICON MONOLITHIC
INTEGRATED CIRCUIT

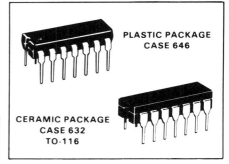

PLASTIC PACKAGE
CASE 646

CERAMIC PACKAGE
CASE 632
TO-116

MAXIMUM RATINGS

Rating	Symbol	Value	Unit
Power Supply Voltage LM139, A LM239, A LM339A LM2901	V_{CC}	+36 or ±18	Vdc
MC3302		+30 or ±15	
Input Differential Voltage Range LM139, A LM239, A LM339, A LM2901	V_{IDR}	36	Vdc
MC3302		30	
Input Common Mode Voltage Range	V_{ICR}	-0 3 to V_{CC}	Vdc
Output Short-Circuit to Gnd (Note 1)	I_{SC}	Continuous	
Input Current ($V_{in} < -0 3$ Vdc) (Note 2)	I_{in}	50	mA
Power Dissipation @ $T_A = 25°C$	P_D		
Ceramic Package		1 0	Watts
Derate above 25°C		80	mW °C
Plastic Package		1 0	Watts
Derate above 25°C		80	mW °C
Operating Ambient Temperature Range	T_A		°C
LM139, A		-55 to +125	
LM239, A		-25 to +85	
LM2901 MC3302		-25 to +85	
LM339, A		0 to +70	
Storage Temperature Range	T_{stg}	-65 to +150	°C

PIN CONNECTIONS

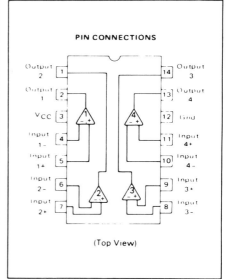

(Top View)

FIGURE 1 — CIRCUIT SCHEMATIC (Diagram shown is for 1 comparator)

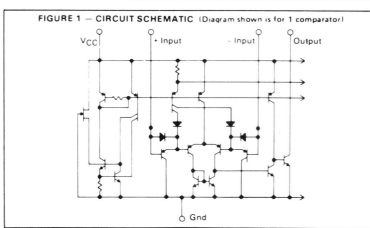

ORDERING INFORMATION

Device	Temperature Range	Package
LM139J, AJ	-55°C to +125°C	Ceramic DIP
LM239J, AJ LM239N, AN	-25°C to +85°C	Ceramic DIP Plastic DIP
LM339J, AJ LM339N, AN	0°C to +70°C	Ceramic DIP Plastic DIP
LM2901N MC3302L MC3302P	-40°C to +85°C	Plastic DIP Ceramic DIP Plastic DIP

© MOTOROLA INC 1982

DS9579

ELECTRICAL CHARACTERISTICS (VCC = +5.0 Vdc, TA = 25°C unless otherwise noted)

Characteristic	Symbol	LM139A			LM239A/339A			LM139			LM239/339			LM2901			MC3302			Unit
		Min	Typ	Max	Min	Typ	Max	Min	Typ	Max	Min	Typ	Max	Min	Typ	Max	Min	Typ	Max	
Input Offset Voltage (Note 4)	VIO	—	±1.0	±2.0	—	±1.0	±2.0	—	±2.0	±5.0	—	±2.0	±5.0	—	±2.0	±7.0	—	±3.0	±20	mVdc
Input Bias Current (Notes 4, 5) (Output in Linear Range)	IIB	—	25	100	—	25	250	—	25	100	—	25	250	—	25	250	—	25	500	nA
Input Offset Current (Note 4)	IIO	—	±3.0	±25	—	±5.0	±50	—	±3.0	±25	—	±5.0	±50	—	±5.0	±50	—	±3.0	±100	nA
Input Common Mode Voltage Range (Note 7)	VICR	0	—	VCC-1.5	0	—	VCC-1.5	0	—	VCC-1.5	0	—	VCC-1.5	0	—	VCC-1.5	0	—	VCC-1.5	V
Supply Current RL = ∞ (For All Comparators) RL = ∞, VCC = 30 Vdc	ICC	— —	0.8 —	2.0 —	— —	0.8 —	2.0 —	— —	0.8 —	2.0 —	— —	0.8 —	2.0 —	— —	0.8 1.0	2.0 2.5	— —	0.8 —	2.0 —	mA
Voltage Gain RL ≥ 15 kΩ, VCC = 15 Vdc	AV	50	200	—	50	200	—	25	200	—	25	200	—	25	100	—	2	30	—	V/mV
Large Signal Response Time VI = TTL Logic Swing, Vref = 1.4 Vdc, VRL = 5.0 Vdc, RL = 5.1 kΩ		—	300	—	—	300	—	—	300	—	—	300	—	—	300	—	—	300	—	ns
Response Time (Note 6) VRL = 5.0 Vdc, RL = 5.1 kΩ		—	1.3	—	—	1.3	—	—	1.3	—	—	1.3	—	—	1.3	—	—	1.3	—	µs
Output Sink Current VI(-) ≥ +1.0 Vdc, VI(+) = 0, VO ≤ 1.5 Vdc	Isink	6.0	16	—	6.0	16	—	6.0	16	—	6.0	16	—	6.0	16	—	6.0	16	—	mA
Saturation Voltage VI(-) ≥ +1.0 Vdc, VI(+) = 0, Isink ≤ 4.0 mA	Vsat	—	130	400	—	130	400	—	130	400	—	130	400	—	130	400	—	130	500	mV
Output Leakage Current VI(+) ≥ +1.0 Vdc, VI(-) = 0, VO = +5.0 Vdc	IOL	—	0.1	—	—	0.1	—	—	0.1	—	—	0.1	—	—	0.1	—	—	0.1	—	nA

PERFORMANCE CHARACTERISTICS (VCC = +5.0 Vdc, TA = Tlow to Thigh (Note 3))

Characteristic	Symbol	LM139A		LM239A/339A		LM139		LM239/339		LM2901		MC3302		Unit
		Min	Max	Min	Max	Min	Max	Min	Max	Min	Max	Min	Max	
Input Offset Voltage (Note 4)	VIO	—	±4.0	—	±4.0	—	±9.0	—	±9.0	—	±15	—	±40	mVdc
Input Bias Current (Notes 4, 5) (Output in Linear Range)	IIB	—	300	—	400	—	300	—	400	—	500	—	1000	nA
Input Offset Current (Note 4)	IIO	—	±100	—	±150	—	±100	—	±150	—	±200	—	±300	nA
Input Common-Mode Voltage Range	VICR	0	VCC-2.0	0	VCC-2.0	0	VCC-2.0	0	VCC-2.0	0	VCC-2.0	0	VCC-2.0	V
Saturation Voltage VI(-) ≥ +1.0 Vdc, VI(+) = 0, Isink ≤ 4.0 mA	Vsat	—	700	—	700	—	700	—	700	—	700	—	700	mV
Output Leakage Current VI(+) ≥ +1.0 Vdc, VI(-) = 0, VO = 30 Vdc	IOL	—	1.0	—	1.0	—	1.0	—	1.0	—	1.0	—	1.0	µA
Differential Input Voltage All VI ≥ 0 Vdc (Note 7)	VID	—	VCC	—	VCC	—	VCC	—	VCC	—	VCC	—	VCC	Vdc

NOTES

1. The maximum output current may be as high as 20 mA, independent of the magnitude of VCC. Output short circuits to VCC can cause excessive heating and eventual destruction.

2. This magnitude of input current will only occur if the leads are driven more negative than ground or the negative supply voltage. This is due to the input PNP collector-base junction becoming forward biased, acting as an input clamp diode. There is also a lateral PNP parasitic transistor action which can cause the output voltage of the comparators to go to the VCC voltage level (or ground if overdrive is large) during the time that an input is driven negative. This will not destroy the device when limited to the max rating and normal output states will recover when the inputs become ≥ ground or negative supply.

3. LM139/139A — Tlow = -55°C, Thigh = +125°C   LM339/339A — Tlow = 0°C, Thigh = +70°C
   LM239/239A — Tlow = -25°C, Thigh = +85°C   LM2901/MC3302 — Tlow = -40°C, Thigh = +85°C

4. At the output switch point, VO = 1.4 Vdc, RS = 100 Ω, 5.0 Vdc ≤ VCC ≤ 30 Vdc. with the inputs over the full common-mode range (0 Vdc to VCC - 1.5 Vdc)

5. The bias current flows out of the inputs due to the PNP input stage. This current is virtually constant, independent of the output state.

6. The response time specified is for a 100 mV input step with 5.0 mV overdrive. For larger signals, 300 ns is typical.

7. Positive excursions of input voltage may exceed the power supply level. As long as one of the inputs remain within the common-mode range, the comparator will provide the proper output state. With VCC = 5.0 Vdc, VI should be limited to 25 volts max. Limiting resistors should be used on all inputs that might exceed VCC.

MOTOROLA *Semiconductor Products Inc.*

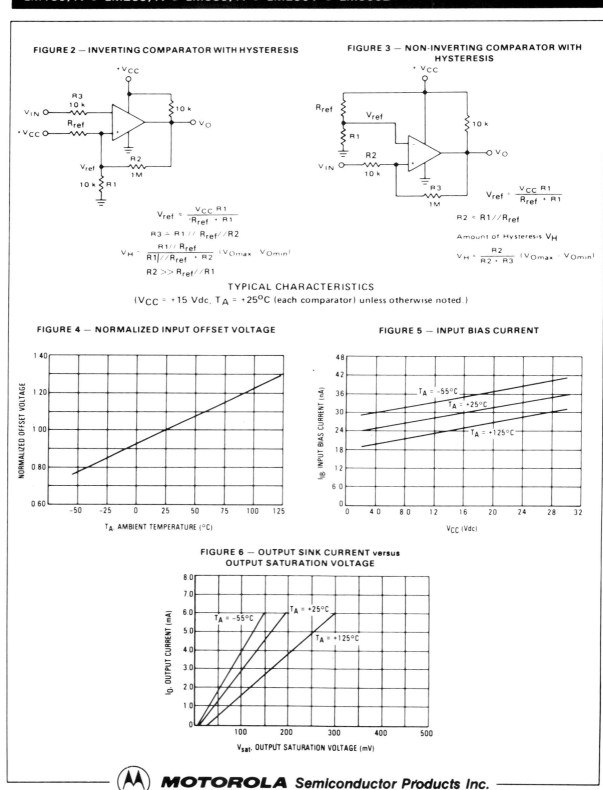

FIGURE 2 — INVERTING COMPARATOR WITH HYSTERESIS

FIGURE 3 — NON-INVERTING COMPARATOR WITH HYSTERESIS

$$V_{ref} = \frac{V_{CC} R1}{R_{ref} + R1}$$

$$R3 = R1 // R_{ref} // R2$$

$$V_H = \frac{R1 // R_{ref}}{R1 // R_{ref} + R2} (V_{Omax} - V_{Omin})$$

$$R2 >> R_{ref} // R1$$

$$V_{ref} = \frac{V_{CC} R1}{R_{ref} + R1}$$

$$R2 = R1 // R_{ref}$$

Amount of Hysteresis V_H

$$V_H = \frac{R2}{R2 + R3} (V_{Omax} - V_{Omin})$$

TYPICAL CHARACTERISTICS

(V_{CC} = +15 Vdc, T_A = +25°C (each comparator) unless otherwise noted.)

FIGURE 4 — NORMALIZED INPUT OFFSET VOLTAGE

FIGURE 5 — INPUT BIAS CURRENT

FIGURE 6 — OUTPUT SINK CURRENT versus OUTPUT SATURATION VOLTAGE

MOTOROLA *Semiconductor Products Inc.*

FIGURE 7 — DRIVING LOGIC

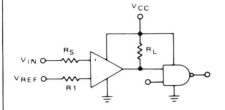

R_S = Source Resistance
$R1 \sim R_S$

LOGIC	DEVICE	V_{CC} Volts	R_L kΩ
CMOS	1/4 MC14001	+15	100
TTL	1/4 MC7400	+5	10

FIGURE 8 — SQUAREWAVE OSCILLATOR

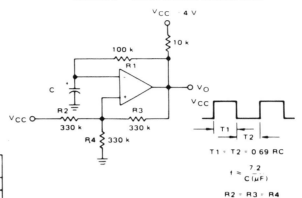

$T1 = T2 = 0.69$ RC

$f \approx \dfrac{7.2}{C\,(\mu F)}$

$R2 = R3 = R4$

$R1 \approx R2//R3//R4$

APPLICATIONS INFORMATION

These quad comparators feature high gain, wide bandwidth characteristics. This gives the device oscillation tendencies if the outputs are capacitively coupled to the inputs via stray capacitance. This oscillation manifests itself during output transistions (V_{OL} to V_{OH}). To alleviate this situation input resistors <10 kΩ should be used. The addi-

tion of positive feedback (<10 mV) is also recommended.
It is good design practice to ground all unused input pins.
Differential input voltages may be larger than supply voltage without damaging the comparator's input voltages. More negative than -300 mV should not be used.

FIGURE 9 — ZERO CROSSING DETECTOR
(Single Supply)

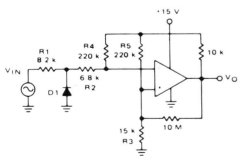

D1 prevents input from going negative by more than 0.6 V

$R1 + R2 = R3$

$R3 \leqslant \dfrac{R5}{10}$ for small error in zero crossing

FIGURE 10 — ZERO CROSSING DETECTOR
(Split Supplies)

$V_{INmin} \approx 0.4$ V peak for 1% phase distortion ((-))

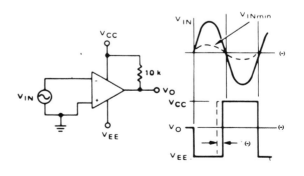

 MOTOROLA *Semiconductor Products Inc.*

OUTLINE DIMENSIONS

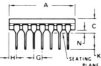

L SUFFIX
CERAMIC PACKAGE
CASE 632
TO-116

NOTES
1. ALL RULES AND NOTES ASSOCIATED
   WITH MO-001 AA OUTLINE SHALL APPLY
2. DIMENSION "L" TO CENTER OF LEADS
   WHEN FORMED PARALLEL
3. LEADS WITHIN 0.25mm (0.010) DIA OF TRUE
   POSITION AT SEATING PLANE AND MAXIMUM
   MATERIAL CONDITION

DIM	MILLIMETERS		INCHES	
	MIN	MAX	MIN	MAX
A	16.8	19.9	0.660	0.785
B	5.59	7.11	0.220	0.280
C	–	5.08	–	0.200
D	0.381	0.584	0.015	0.023
F	0.77	1.77	0.030	0.070
G	2.54 BSC		0.100 BSC	
J	0.203	0.381	0.008	0.015
K	2.54	–	0.100	–
L	7.62 BSC		0.300 BSC	
M	–	15°	–	15°
N	0.51	0.76	0.020	0.030
P	–	8.25	–	0.325

All JEDEC dimensions and notes apply

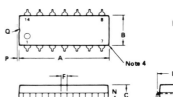

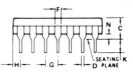

P SUFFIX
PLASTIC PACKAGE
CASE 646

Note 4

NOTES
1. LEADS WITHIN 0.13 mm
   (0.005) RADIUS OF TRUE
   POSITION AT SEATING
   PLANE AT MAXIMUM
   MATERIAL CONDITION.
2. DIMENSION "L" TO
   CENTER OF LEADS
   WHEN FORMED
   PARALLEL
3. DIMENSION "B" DOES NOT
   INCLUDE MOLD FLASH
4. ROUNDED CORNERS OPTIONAL

DIM	MILLIMETERS		INCHES	
	MIN	MAX	MIN	MAX
A	18.16	19.56	0.715	0.770
B	6.10	6.60	0.240	0.260
C	4.06	5.08	0.160	0.200
D	0.38	0.53	0.015	0.021
F	1.02	1.78	0.040	0.070
G	2.54 BSC		0.100 BSC	
H	1.32	2.41	0.052	0.095
J	0.20	0.38	0.008	0.015
K	2.92	3.43	0.115	0.135
L	7.62 BSC		0.300 BSC	
M	0°	10°	0°	10°
N	0.51	1.02	0.020	0.040

Motorola reserves the right to make changes to any products herein to improve reliability, function or design. Motorola does not assume any liability arising out of the application or use of any product or circuit described herein; neither does it convey any license under its patent rights nor the rights of others.

 MOTOROLA *Semiconductor Products Inc.*

MC1741, MC1741C
MC1741N, MC1741NC

OPERATIONAL AMPLIFIER
SILICON MONOLITHIC
INTEGRATED CIRCUIT

INTERNALLY COMPENSATED, HIGH PERFORMANCE OPERATIONAL AMPLIFIERS

. . . designed for use as a summing amplifier, integrator, or amplifier with operating characteristics as a function of the external feedback components.

- No Frequency Compensation Required
- Short-Circuit Protection
- Offset Voltage Null Capability
- Wide Common-Mode and Differential Voltage Ranges
- Low-Power Consumption
- No Latch Up
- Low Noise Selections Offered — N Suffix

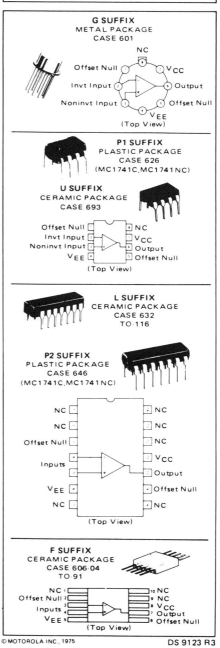

MAXIMUM RATINGS ($T_A = +25^\circ C$ unless otherwise noted)

Rating	Symbol	MC1741C	MC1741	Unit
Power Supply Voltage	V_{CC}	+18	+22	Vdc
	V_{EE}	-18	-22	Vdc
Input Differential Voltage	V_{ID}	±30		Volts
Input Common Mode Voltage (Note 1)	V_{ICM}	+15		Volts
Output Short Circuit Duration (Note 2)	t_S	Continuous		
Operating Ambient Temperature Range	T_A	0 to +70	-55 to +125	$^\circ C$
Storage Temperature Range	T_{stg}			$^\circ C$
Metal, Flat and Ceramic Packages		-65 to +150		
Plastic Packages		-55 to +125		

Note 1. For supply voltages less than + 15 V, the absolute maximum input voltage is equal to the supply voltage.

Note 2. Supply voltage equal to or less than 15 V.

EQUIVALENT CIRCUIT SCHEMATIC

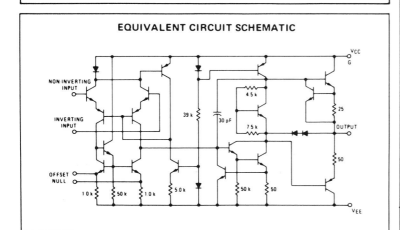

© MOTOROLA INC., 1975

DS 9123 R3

ELECTRICAL CHARACTERISTICS (V_{CC} = 15 V, V_{EE} = 15 V, T_A = 25°C unless otherwise noted).

Characteristic	Symbol	MC1741 Min	MC1741 Typ	MC1741 Max	MC1741C Min	MC1741C Typ	MC1741C Max	Unit
Input Offset Voltage ($R_S \leqslant$ 10 k)	V_{IO}	–	1.0	5.0	–	2.0	6.0	mV
Input Offset Current	I_{IO}	--	20	200	–	20	200	nA
Input Bias Current	I_{IB}	–	80	500	–	80	500	nA
Input Resistance	r_i	0.3	2.0	--	0.3	2.0	--	MΩ
Input Capacitance	C_i	–	1.4	–	–	1.4	–	pF
Offset Voltage Adjustment Range	V_{IOR}	-	±15	–	–	±15	–	mV
Common Mode Input Voltage Range	V_{ICR}	±12.	±13	–	±12	±13	–	V
Large Signal Voltage Gain (V_O = ±10 V, $R_L \geqslant$ 2.0 k)	A_V	50	200	–	20	200	–	V/mV
Output Resistance	r_o	–	75	–	–	75	–	Ω
Common Mode Rejection Ratio ($R_S \leqslant$ 10 k)	CMRR	70	90	–	70	90	–	dB
Supply Voltage Rejection Ratio ($R_S \leqslant$ 10 k)	PSRR	–	30	150	–	30	150	μV/V
Output Voltage Swing ($R_L \geqslant$ 10 k) ($R_L \geqslant$ 2 k)	V_O	±12 ±10	±14 ±13	· –	±12 ±10	±14 ±13	-- –	V
Output Short-Circuit Current	I_{os}	–	20	–	–	20	--	mA
Supply Current	I_D	-	1.7	2.8	–	1.7	2.8	mA
Power Consumption	P_C	–	50	85	-- ·	50	85	mW
Transient Response (Unity Gain – Non-Inverting) (V_I = 20 mV, $R_L \geqslant$ 2 k, $C_L \leqslant$ 100 pF) Rise Time	t_{TLH}	·	0.3	·	–	0.3	·	μs
(V_I = 20 mV, $R_L \geqslant$ 2 k, $C_L \leqslant$ 100 pF) Overshoot	os	--	15	·	–	15	·	%
(V_I = 10 V, $R_L \geqslant$ 2 k, $C_L \leqslant$ 100 pF) Slew Rate	SR	–	0.5	–	–	0.5	–	V/μs

ELECTRICAL CHARACTERISTICS (V_{CC} = 15 V, V_{EE} = 15 V, T_A = *T_{high} to T_{low} unless otherwise noted.)

Characteristic	Symbol	MC1741 Min	MC1741 Typ	MC1741 Max	MC1741C Min	MC1741C Typ	MC1741C Max	Unit
Input Offset Voltage ($R_S \leqslant$ 10 kΩ)	V_{IO}	–	1.0	6.0	–	–	7.5	mV
Input Offset Current (T_A = 125°C) (T_A = -55°C) (T_A = 0°C to +70°C)	I_{IO}	- –	7.0 85 –	200 500 –	– –	– · --	– – 300	nA
Input Bias Current (T_A = 125°C) (T_A = -55°C) (T_A = 0°C to +70°C)	I_{IB}	– -- --	30 300 –	500 1500 –	– – –	– – --	– – 800	nA
Common Mode Input Voltage Range	V_{ICR}	±12	+13	–	–	–	–	V
Common Mode Rejection Ratio ($R_S \leqslant$ 10 k)	CMRR	70	90	–	–	--	–	dB
Supply Voltage Rejection Ratio ($R_S \leqslant$ 10 k)	PSRR	–	30	150	–	–	–	μV/V
Output Voltage Swing ($R_L \geqslant$ 10 k) ($R_L \geqslant$ 2 k)	V_O	±12 ±10	±14 ±13	– –	-- ±10	- ±13	– –	V
Large Signal Voltage Gain ($R_L \geqslant$ 2 k, V_{out} = ±10 V)	A_V	25	–	–	15	·	·	V/mV
Supply Currents (T_A = 125°C) (T_A = -55°C)	I_D	– --	1.5 2.0	2.5 3.3			;	mA
Power Consumption (T_A = +125°C) (T_A = -55°C)	P_C	– ·	45 60	75 100				mW

*T_{high} = 125°C for MC1741 and 70°C for MC1741C
T_{low} = -55°C for MC1741 and 0°C for MC1741C

MOTOROLA *Semiconductor Products Inc.*

NOISE CHARACTERISTICS (Applies for MC1741N and MC1741NC only, V_{CC} = 15 V, V_{EE} = –15 V, T_A = +25°C)

Characteristic	Symbol	MC1741N			MC1741NC			Unit
		Min	Typ	Max	Min	Typ	Max	
Burst Noise (Popcorn Noise) (BW = 1.0 Hz to 1.0 kHz, t = 10 s, R_S = 100 k) (Input Referenced)	E_n	–	–	20	–	–	20	μV/peak

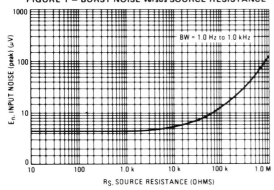

FIGURE 1 – BURST NOISE versus SOURCE RESISTANCE

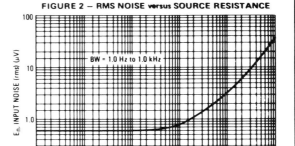

FIGURE 2 – RMS NOISE versus SOURCE RESISTANCE

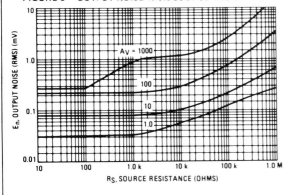

FIGURE 3 – OUTPUT NOISE versus SOURCE RESISTANCE

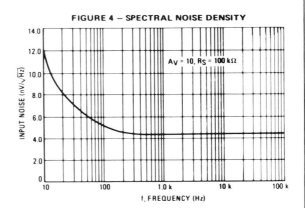

FIGURE 4 – SPECTRAL NOISE DENSITY

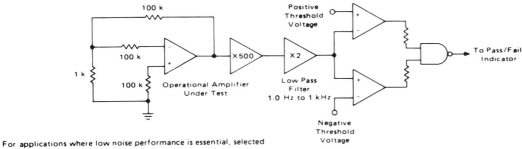

FIGURE 5 – BURST NOISE TEST CIRCUIT (N Suffixed Devices Only)

For applications where low noise performance is essential, selected devices denoted by an N suffix are offered. These units have been 100% tested for burst noise pulses on a special noise test system. Unlike conventional peak reading or RMS meters, this system was especially designed to provide the quick response time essential to burst (popcorn) noise testing.

The test time employed is 10 seconds and the 20 μV peak limit refers to the operational amplifier input thus eliminating errors in the closed-loop gain factor of the operational amplifier under test.

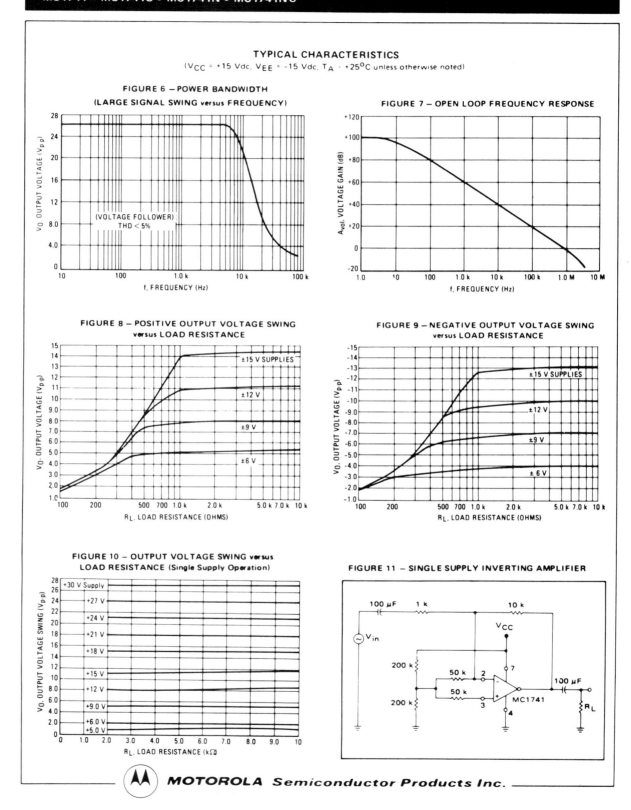

TYPICAL CHARACTERISTICS

$(V_{CC} = +15$ Vdc, $V_{EE} = -15$ Vdc, $T_A = +25°C$ unless otherwise noted)

FIGURE 6 – POWER BANDWIDTH
(LARGE SIGNAL SWING versus FREQUENCY)

FIGURE 7 – OPEN LOOP FREQUENCY RESPONSE

FIGURE 8 – POSITIVE OUTPUT VOLTAGE SWING versus LOAD RESISTANCE

FIGURE 9 – NEGATIVE OUTPUT VOLTAGE SWING versus LOAD RESISTANCE

FIGURE 10 – OUTPUT VOLTAGE SWING versus LOAD RESISTANCE (Single Supply Operation)

FIGURE 11 – SINGLE SUPPLY INVERTING AMPLIFIER

MOTOROLA *Semiconductor Products Inc.*

FIGURE 12 – NON-INVERTING PULSE RESPONSE

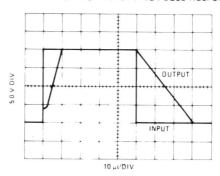

FIGURE 13 – TRANSIENT RESPONSE TEST CIRCUIT

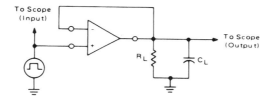

FIGURE 14 – OPEN LOOP VOLTAGE GAIN
versus SUPPLY VOLTAGE

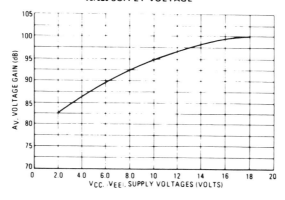

Circuit diagrams utilizing Motorola products are included as a means of illustrating typical semiconductor applications, consequently, complete information sufficient for construction purposes is not necessarily given. The information has been carefully checked and is believed to be entirely reliable. However, no responsibility is assumed for inaccuracies. Furthermore, such information does not convey to the purchaser of the semiconductor devices described any license under the patent rights of Motorola Inc. or others.

 MOTOROLA *Semiconductor Products Inc.*

THERMAL INFORMATION

The maximum power consumption an integrated circuit can tolerate at a given operating ambient temperature, can be found from the equation:

$$P_{D(T_A)} = \frac{T_{J(max)} - T_A}{R_{\theta JA}(Typ)}$$

Where: $P_{D(T_A)}$ = Power Dissipation allowable at a given operating ambient temperature. This must be greater than

the sum of the products of the supply voltages and supply currents at the worst case operating condition.

$T_{J(max)}$ = Maximum Operating Junction Temperature as listed in the Maximum Ratings Section

T_A = Maximum Desired Operating Ambient Temperature

$R_{\theta JA}(Typ)$ = Typical Thermal Resistance Junction to Ambient

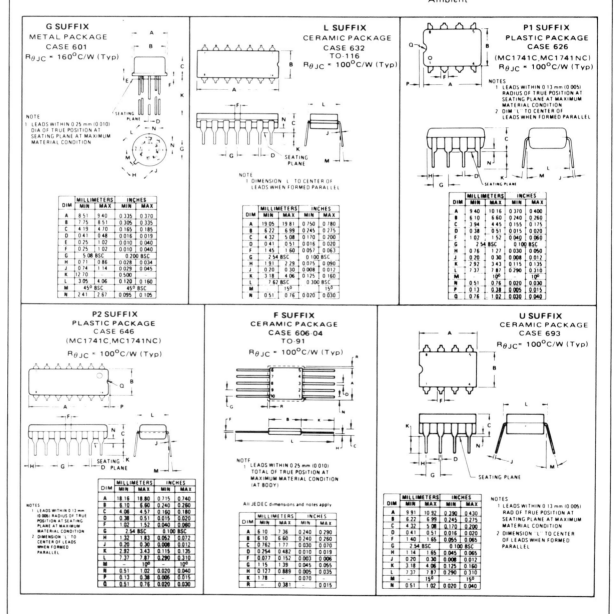

G SUFFIX METAL PACKAGE CASE 601 $R_{\theta JC}$ = 160°C/W (Typ)

NOTE
1 LEADS WITHIN 0 25 mm (0 010) DIA OF TRUE POSITION AT SEATING PLANE AT MAXIMUM MATERIAL CONDITION

DIM	MILLIMETERS		INCHES	
	MIN	MAX	MIN	MAX
A	8 51	9 40	0.335	0.370
B	7 75	8 51	0.305	0.335
C	4 19	4 70	0.165	0.185
D	0 41	0 48	0.016	0.019
E	0 25	1 02	0.010	0.040
F	0 25	1 02	0.010	0.040
G	5 08 BSC		0.200 BSC	
H	0 71	0 86	0.028	0.034
J	0 74	1 14	0.029	0.045
K	12 70	-	0.500	-
L	3 05	4 06	0.120	0.160
M	45° BSC		45° BSC	
N	2 41	2 67	0.095	0.105

L SUFFIX CERAMIC PACKAGE CASE 632 TO-116 $R_{\theta JC}$ = 100°C/W (Typ)

NOTE
1 DIMENSION L TO CENTER OF LEADS WHEN FORMED PARALLEL

DIM	MILLIMETERS		INCHES	
	MIN	MAX	MIN	MAX
A	19 05	19 81	0.750	0.780
B	6 22	6 99	0.245	0.275
C	4 32	5 08	0.170	0.200
D	0 41	0 51	0.016	0.020
F	1 45	1 60	0.057	0.063
G	2 54 BSC		0 100 BSC	
H	1 91	2 29	0.075	0.090
J	0.20	0 30	0.008	0.012
K	3 18	4 06	0.125	0.160
L	7 62 BSC		0 300 BSC	
M	15°		15°	
N	0 51	0 76	0.020	0.030

P1 SUFFIX PLASTIC PACKAGE CASE 626 (MC1741C,MC1741NC) $R_{\theta JC}$ = 100°C/W (Typ)

NOTES
1 LEADS WITHIN 0 13 mm (0 005) RADIUS OF TRUE POSITION AT SEATING PLANE AT MAXIMUM MATERIAL CONDITION
2 DIM L TO CENTER OF LEADS WHEN FORMED PARALLEL

DIM	MILLIMETERS		INCHES	
	MIN	MAX	MIN	MAX
A	9 40	10 16	0.370	0.400
B	6 10	6 60	0.240	0.260
C	3 94	4 45	0.155	0.175
D	0 38	0 51	0.015	0.020
F	1 02	1 52	0.040	0.060
G	2 54 BSC		0 100 BSC	
H	0 76	1 27	0.030	0.050
J	0 20	0 30	0.008	0.012
K	2 92	3 43	0.115	0.135
L	7 37	7 87	0.290	0.310
M		10°		10°
N	0 51	0 76	0.020	0.030
P	0 13	0 38	0.005	0.015
Q	0 76	1 02	0.030	0.040

P2 SUFFIX PLASTIC PACKAGE CASE 646 (MC1741C,MC1741NC) $R_{\theta JC}$ = 100°C/W (Typ)

NOTES
1 LEADS WITHIN 0 13 mm (0 005) RADIUS OF TRUE POSITION AT SEATING PLANE AT MAXIMUM MATERIAL CONDITION
2 DIMENSION L TO CENTER OF LEADS WHEN FORMED PARALLEL

DIM	MILLIMETERS		INCHES	
	MIN	MAX	MIN	MAX
A	18 16	18 80	0.715	0.740
B	6 10	6 60	0.240	0.260
C	4 06	4 57	0.160	0.180
D	0 38	0 51	0.015	0.020
F	1 02	1 52	0.040	0.060
G	2 54 BSC		0 100 BSC	
H	1 32	1 83	0.052	0.072
J	0 20	0 30	0.008	0.012
K	2 92	3 43	0.115	0.135
L	7 37	7 87	0.290	0.310
M		10°		10°
N	0 51	1 02	0.020	0.040
P	0 13	0 38	0.005	0.015
Q	0 51	0 76	0.020	0.030

F SUFFIX CERAMIC PACKAGE CASE 606-04 TO-91 $R_{\theta JC}$ = 100°C/W (Typ)

NOTE
1 LEADS WITHIN 0 25 mm (0 010) TOTAL OF TRUE POSITION AT MAXIMUM MATERIAL CONDITION (AT BODY)

All JEDEC dimensions and notes apply

DIM	MILLIMETERS		INCHES	
	MIN	MAX	MIN	MAX
A	6 10	7 36	0.240	0.290
B	6 10	6 60	0.240	0.260
C	0 762	1 77	0.030	0.070
D	0 254	0 482	0.010	0.019
F	0.077	0 152	0.003	0.006
G	1 15	1 39	0.045	0.055
H	0 127	0 889	0.005	0.035
K	1 78	-	0.070	-
R	-	0 381	-	0 015

U SUFFIX CERAMIC PACKAGE CASE 693 $R_{\theta JC}$ = 100°C/W (Typ)

NOTES
1 LEADS WITHIN 0 13 mm (0 005) RAD OF TRUE POSITION AT SEATING PLANE AT MAXIMUM MATERIAL CONDITION
2 DIMENSION L TO CENTER OF LEADS WHEN FORMED PARALLEL

DIM	MILLIMETERS		INCHES	
	MIN	MAX	MIN	MAX
A	9 91	10 92	0.390	0.430
B	6 22	6 99	0.245	0.275
C	4 32	5 08	0.170	0.200
D	0 41	0 51	0.016	0.020
F	1 40	1 65	0.055	0.065
G	2 54 BSC		0 100 BSC	
H	1 14	1 65	0.045	0.065
J	0 20	0 30	0.008	0.012
K	3 18	4 06	0.125	0.160
L	7 37	7 87	0.290	0.310
M		15°		15°
N	0 51	1 02	0.020	0.040

MOTOROLA *Semiconductor Products Inc.*

BOX 20912 ● PHOENIX, ARIZONA 85036 ● A SUBSIDIARY OF MOTOROLA INC

4560-11 PRINTED IN USA 12-75 IMPERIAL LITHO 54476   ICH

05-9123-R3

MOTOROLA
Semiconductors
BOX 20912 • PHOENIX, ARIZONA 85036

PNP	NPN
MJE170	**MJE180**
MJE171	**MJE181**
MJE172	**MJE182**

COMPLEMENTARY PLASTIC SILICON POWER TRANSISTORS

. . . designed for low power audio amplifier and low current, high speed switching applications.

- Collector-Emitter Sustaining Voltage —
  $V_{CEO(sus)}$ = 40 Vdc — MJE170, MJE180
  = 60 Vdc — MJE171, MJE181
  = 80 Vdc — MJE172, MJE182

- DC Current Gain —
  h_{FE} = 30 (Min) @ I_C = 0.5 Adc
  = 12 (Min) @ I_C = 1.5 Adc

- Current-Gain — Bandwidth Product —
  f_T = 50 MHz (Min) @ I_C = 100 mAdc

- Annular Construction for Low Leakages —
  I_{CBO} = 100 nA (Max) @ Rated V_{CB}

3 AMPERE POWER TRANSISTORS COMPLEMENTARY SILICON
40-60-80 VOLTS
12.5 WATTS

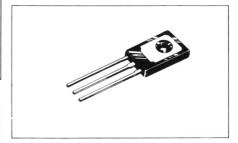

MAXIMUM RATINGS

Rating	Symbol	MJE170 MJE180	MJE171 MJE181	MJE172 MJE182	Unit
Collector-Base Voltage	V_{CB}	60	80	100	Vdc
Collector-Emitter Voltage	V_{CEO}	40	60	80	Vdc
Emitter-Base Voltage	V_{EB}	← 7.0 →			Vdc
Collector Current — Continuous Peak	I_C	← 3.0 → ← 6.0 →			Adc
Base Current	I_B	← 1.0 →			Adc
Total Device Dissipation @ T_A = 25°C Derate above 25°C	P_D	← 1.5 → ← 0.012 →			Watts W/°C
Total Device Dissipation @ T_C = 25°C Derate above 25°C	P_D	← 12.5 → ← 0.1 →			Watts W/°C
Operating and Storage Junction Temperature Range	T_J, T_{stg}	← −65 to +150 →			°C

THERMAL CHARACTERISTICS

Characteristic	Symbol	Max	Unit
Thermal Resistance, Junction to Case	θ_{JC}	10	°C/W
Thermal Resistance, Junction to Ambient	θ_{JA}	83.4	°C/W

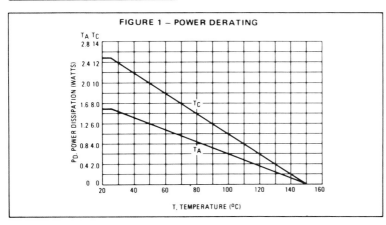

FIGURE 1 — POWER DERATING

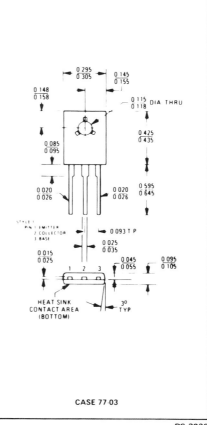

CASE 77-03

DS 3239

ELECTRICAL CHARACTERISTICS ($T_C = 25^\circ C$ unless otherwise noted)

Characteristic	Symbol	Min	Max	Unit
OFF CHARACTERISTICS				
Collector-Emitter Sustaining Voltage	$V_{CEO(sus)}$		—	Vdc
($I_C = 10$ mAdc, $I_B = 0$) MJE170, MJE180		40	—	
MJE171, MJE181		60	—	
MJE172, MJE182		80	—	
Collector Cutoff Current	I_{CBO}			μAdc
($V_{CB} = 60$ Vdc, $I_E = 0$) MJE170, MJE180		—	0.1	
($V_{CB} = 80$ Vdc, $I_E = 0$) MJE171, MJE181		—	0.1	
($V_{CB} = 100$ Vdc, $I_E = 0$) MJE172, MJE182		—	0.1	
($V_{CB} = 60$ Vdc, $I_E = 0$, $T_C = 150^\circ C$) MJE170, MJE180		—	0.1	mAdc
($V_{CB} = 80$ Vdc, $I_E = 0$, $T_C = 150^\circ C$) MJE171, MJE181		—	0.1	
($V_{CB} = 100$ Vdc, $I_E = 0$, $T_C = 150^\circ C$) MJE172, MJE182		—	0.1	
Emitter Cutoff Current	I_{EBO}			μAdc
($V_{BE} = 7.0$ Vdc, $I_C = 0$)		—	0.1	
ON CHARACTERISTICS				
DC Current Gain	h_{FE}			—
($I_C = 100$ mAdc, $V_{CE} = 1.0$ Vdc)		50	250	
($I_C = 500$ mAdc, $V_{CE} = 1.0$ Vdc)		30	—	
($I_C = 1.5$ Adc, $V_{CE} = 1.0$ Vdc)		12	—	
Collector-Emitter Saturation Voltage	$V_{CE(sat)}$			Vdc
($I_C = 500$ mAdc, $I_B = 50$ mAdc)		—	0.3	
($I_C = 1.5$ Adc, $I_B = 150$ mAdc)		—	0.9	
($I_C = 3.0$ Adc, $I_B = 600$ mAdc)		—	1.7	
Base-Emitter Saturation Voltage	$V_{BE(sat)}$			Vdc
($I_C = 1.5$ Adc, $I_B = 150$ mAdc)		—	1.5	
($I_C = 3.0$ Adc, $I_B = 600$ mAdc)		—	2.0	
Base-Emitter On Voltage	$V_{BE(on)}$			Vdc
($I_C = 500$ mAdc, $V_{CE} = 1.0$ Vdc)		—	1.2	
DYNAMIC CHARACTERISTICS				
Current-Gain — Bandwidth Product (1)	f_T			MHz
($I_C = 100$ mAdc, $V_{CE} = 10$ Vdc, $f_{test} = 10$ MHz)		50	—	
Output Capacitance	C_{ob}			pF
($V_{CB} = 10$ Vdc, $I_E = 0$, $f = 0.1$ MHz) MJE170/MJE172		—	50	
MJE180/MJE182		—	30	

(1) $f_T = |h_{fe}| \bullet f_{test}$

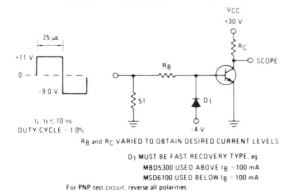

FIGURE 2 — SWITCHING TIME TEST CIRCUIT

$t_r, t_f \leq 10$ ns
DUTY CYCLE = 1.0%

R_B and R_C VARIED TO OBTAIN DESIRED CURRENT LEVELS

D_1 MUST BE FAST RECOVERY TYPE, eg
MBD5300 USED ABOVE $I_B \approx 100$ mA
MSD6100 USED BELOW $I_B \approx 100$ mA
For PNP test circuit, reverse all polarities.

FIGURE 3 — TURN-ON TIME

$V_{CC} = 30$ V
$I_C/I_B = 10$
$V_{BE(off)} = 4.0$ V
$T_J = 25^\circ C$

PNP MJE170/MJE172
NPN MJE180/MJE182

I_C, COLLECTOR CURRENT (AMP)

t, TIME (ns)

MOTOROLA *Semiconductor Products Inc.*

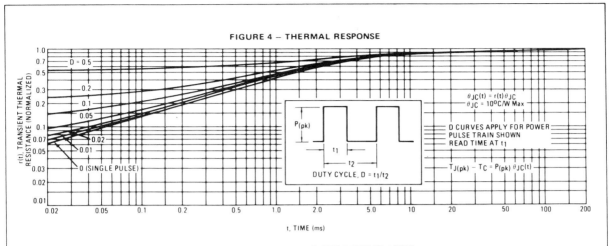

FIGURE 4 – THERMAL RESPONSE

ACTIVE-REGION SAFE OPERATING AREA

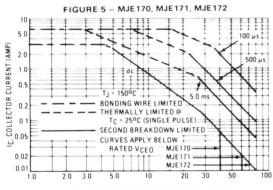

FIGURE 5 – MJE170, MJE171, MJE172

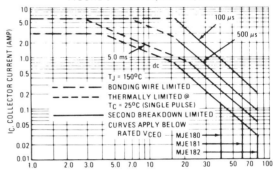

FIGURE 6 – MJE180, MJE181, MJE182

There are two limitations on the power handling ability of a transistor – average junction temperature and second breakdown. Safe operating area curves indicate I_C – V_{CE} limits of the transistor that must be observed for reliable operation; i.e., the transistor must not be subjected to greater dissipation than the curves indicate.

The data of Figures 5 and 6 is based on $T_{J(pk)} = 150°C$; T_C is

variable depending on conditions. Second breakdown pulse limits are valid for duty cycles to 10% provided $T_{J(pk)} < 150°C$. $T_{J(pk)}$ may be calculated from the data in Figure 4. At high case temperature, thermal limitations will reduce the power that can be handled to values less than the limitations imposed by second breakdown. (See AN-415)

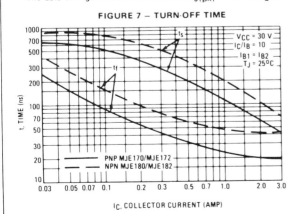

FIGURE 7 – TURN-OFF TIME

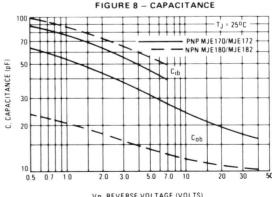

FIGURE 8 – CAPACITANCE

 MOTOROLA *Semiconductor Products Inc.*

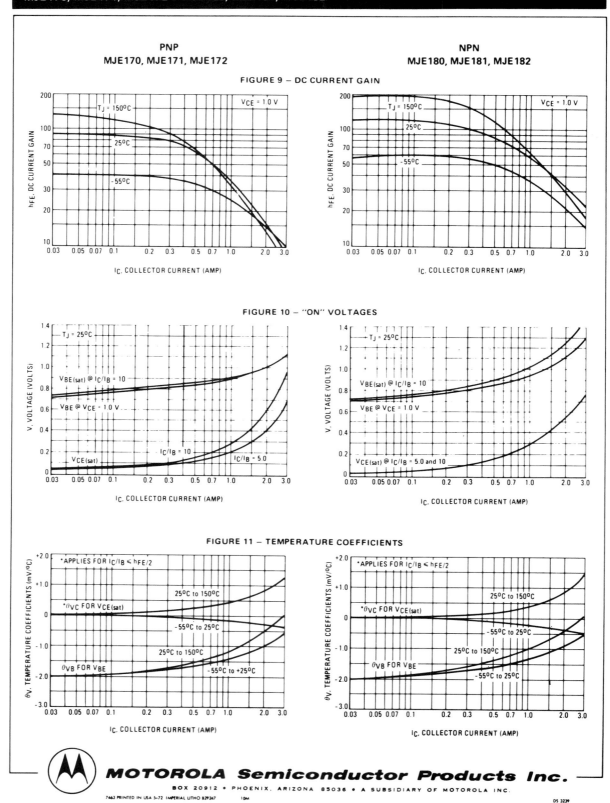

PNP
MJE170, MJE171, MJE172

NPN
MJE180, MJE181, MJE182

FIGURE 9 – DC CURRENT GAIN

FIGURE 10 – "ON" VOLTAGES

FIGURE 11 – TEMPERATURE COEFFICIENTS

MOTOROLA *Semiconductor Products Inc.*

BOX 20912 ● PHOENIX, ARIZONA 85036 ● A SUBSIDIARY OF MOTOROLA INC.

7463 PRINTED IN USA 5-72 IMPERIAL LITHO B29367   10M

DS 3239

3130 (RCA)

CMOS OPERATIONAL AMPLIFIER

This is a CMOS linear operational amplifier with an extremely high input impedance, a 15-megahertz unity-gain bandwidth, and a 10-volt-per-microsecond slew rate.

Details on its use appear in Unit 6. The high input impedance is very useful in sample/hold, pH meter, comparator, and precision rectifier applications.

Low-frequency open-loop gain is 100,000. A compensation capacitor of 30 pF or more is normally added between pins 1 and 8, as shown, for stable operation. Input offset is typically 8 millivolts and is trimmable with the pot as shown.

The available output current is 20 milliamperes in either direction. The device is somewhat noisy and quite sensitive to output capacitive loading.

The optimum input reference is one-half the supply voltage. Inputs can be referenced as far negative as the pin-4 voltage, but input voltages more than 0.3 volt negative with respect to pin 4 will draw input current and can reverse the output sense in comparator circuits.

Total power-supply voltage can range from 5 to 15 volts (±2.5 to ±7.5). A current of 10 milliamperes is typical with a 15-volt supply. The 3140 is a similar device with bipolar output and higher supply voltages.

0320 (HUGHES)

FREQUENCY SYNTHESIZER

This is a divide-by-n counter and a phase detector in a single package. It is used in frequency synthesizers where a number of frequencies are to be generated from a single crystal reference. Up to 1020 different frequencies can be obtained.

Input signals are accepted at pins 15 or 16. Pin 16 accepts 5-kHz to 1-MHz clock signals and provides internal conditioning. Pin 15 accepts signals to 5 MHz (5-volt supply) or 10 MHz (12-volt supply). A divide-by-n output appears on pin 14. This also drives a phase detector that compares the divide-by-n output frequency against the frequency input to pin 18.

Polarity of the phase detector is set by pin 21. If pin 21 is grounded, the phase detector output is low if the pin-14 frequency is low with respect to the pin-18 frequency, and high if the pin-14 frequency is high with respect to the pin-18 frequency. Making pin 21 positive reverses this output sense. Use a grounded pin 21 if increasing the voltage increases the frequency in any vco used with this chip.

The number n can range from 3 to 1023; n is made of two parts, a binary number from 0 to 128 and a binary-coded decimal number from 0 to 1000. Thumbwheel switches are often used to enter the bcd portion of the division number, while the binary inputs are often hard-wired. The binary inputs make frequency offsets particularly easy, and they facilitate code conversions needed to get to channel numbers or channel frequencies.

Total package current at 5 megahertz is 1 milliampere on a +5-volt supply.

4001

QUAD 2-INPUT NOR GATE

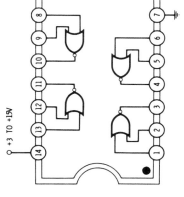

+3 TO +15V

TOP VIEW

All four positive-logic NOR gates may be used independently. On any one gate, with either or both inputs high, the output is low; with both inputs low, the output is high.

Propagation delay is 25 nanoseconds at 10 volts and 60 nanoseconds at 5 volts. Total package current at 1 megahertz is 0.4 milliampere at 5 volts and 0.8 milliampere at 10 volts.

4000

DUAL 3-INPUT NOR GATE PLUS INVERTER

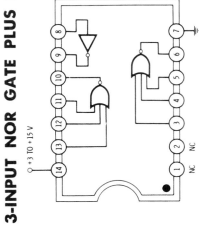

+3 TO +15 V

TOP VIEW

The package contains two 3-input NOR gates and an inverter. They may be used separately.

An input low drives the inverter output high, and vice versa.

On the NOR gates, any input high drives the output low. All inputs low drives the output high.

The gates may be combined. A NOR gate followed by an inverter gives a 3-input OR gate. This output routed to the remaining gate gives a 5-input NOR function.

Propagation delay is 25 nanoseconds at 10 volts and 60 nanoseconds at 5 volts. Total package current is 0.3 mA at 5 volts and 0.6 mA at 10 volts at a 1-megahertz data rate.

4008

4-BIT FULL ADDER

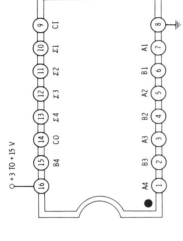

TOP VIEW

This is an arithmetic package that gives the positive-logic sum of two 4-bit binary numbers. Packages may be cascaded for more bits.

Input word A is applied to the A inputs with weights $A1 = 1$, $A2 = 2$, $A3 = 4$, and $A4 = 8$. Input word B is applied to the B inputs with weights $B1 = 1$, $B2 = 2$, $B3 = 4$, and $B4 = 8$. The positive-logic binary sum of the two input words appears on the Σ outputs, with $\Sigma1 = 1$, $\Sigma2 = 2$, $\Sigma3 = 4$, and $\Sigma4 = 8$ weightings. Should a carry result, it will appear on the CO terminal with a weighting of 16.

The carry input should be grounded on the package working on the four least-significant bits of a binary sum. The CO of this package should go to the CI input of the package working on the next four most-significant bits, and so on. Internal look-ahead carry is done to increase operating speed.

Addition time is 900 nanoseconds at 5 volts and 325 nanoseconds at 10 volts, with newer devices being significantly faster. Total package current is 1.6 milliamperes at 5 volts and 3.2 milliamperes at 10 volts at a 1-megahertz word rate.

4007

DUAL CMOS PAIR PLUS INVERTER

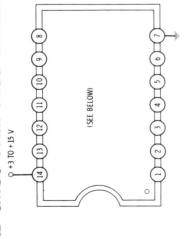

(SEE BELOW)

TOP VIEW

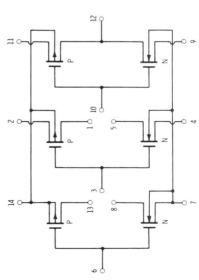

This is basically a "do-it-yourself" kit of CMOS transistors. You can use it for simple logic, transmission gates, buffers, drivers, CMOS variable resistors, discharge transistors for capacitors, analog-to-digital converters, input translators, oscillators, etc. Do not allow any pin voltage to exceed the pin-14 voltage. Do not allow any pin voltage to go below the pin-7 voltage.

Propagation time is 20 nanoseconds at 10 volts and 35 nanoseconds at 5 volts. Total package current is .7 milliampere at 5 volts and 1.4 milliamperes at 10 volts at a 1-megahertz clock rate.

4010

HEX NONINVERTING BUFFER

+3 TO +15 V
BUT MORE THAN
PIN 1 VOLTAGE

NC.

+3 TO +15 V

TOP VIEW

This device should not be used for new designs. Use the 4050 instead.

The 4010 will self-destruct if supply voltages are applied in the wrong sequence. The voltage on pin 16 must always be equal to or less than the voltage on pin 1.

4009

HEX INVERTING BUFFER

+3 TO +15 V
BUT MORE THAN
PIN 1 VOLTAGE

NC

+3 TO +15 V

TOP VIEW

This device should not be used for new designs. Use the 4049 instead.

The 4009 will self-destruct if supply voltages are applied in the wrong sequence. The voltage on pin 16 must always be equal to or less than the voltage on pin 1.

4068

8-INPUT NAND GATE

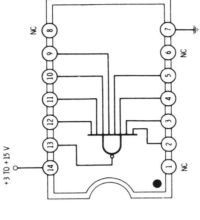

TOP VIEW

This package contains a single positive-logic, 8-input NAND gate.

If one or more inputs are *low*, the output will be *high*. If all eight inputs are *high*, the output will be *low*.

Propagation delay is 130 nanoseconds at 10 volts and 325 nanoseconds at 5 volts. Total package current is .5 milliampere at 5 volts and 1 milliampere at 10 volts.

Note that this is a very slow device. It should not be used in high-speed applications, particularly on a 5-volt or lower supply.

4067

1-OF-16 ANALOG SWITCH

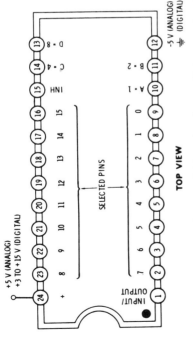

TOP VIEW

This package may be used as a 1-of-16 analog data multiplexer or demultiplexer, or as a 1-of-16 digital selector or distributor.

In the analog mode, -5 is applied to pins 12 and 15 and $+5$ to pin 24.

The channel selected is determined by the A, B, C, and D inputs weighted $A = 1$, $B = 2$, $C = 4$, and $D = 8$, with a zero defined as -5 volts and a 1 defined as $+5$ volts. For instance, with $A = 1$, $B = 0$, $C = 1$, and $D = 1$, channel 13 (pin 18) is connected to the common input/output terminal.

In the digital mode, ground is applied to pins 12 and 15, and $+3$ to $+15$ volts to pin 24. The channel selected is similarly determined, with a zero being ground and a 1 being defined as the pin-24 voltage.

The ON resistance is 200 ohms. On-channel frequency response extends to 40 megahertz. Off-channel cross talk is -40 decibels (1/100 amplitude) for frequencies less than 1 megahertz. The inhibit pin will turn all channels off if brought to the pin-24 voltage. More than one channel can be on during settling times. This can be eliminated by using the Inhibit before, during, and immediately after channel selection.

All inputs and outputs must be less than the pin-24 voltage and more than the pin-12 voltage.

Select and inhibit propagation times are 200 nanoseconds at 10 volts and 400 nanoseconds at 5 volts. Power dissipation depends on the loading and the frequency of operation.

7207
(INTERSIL)

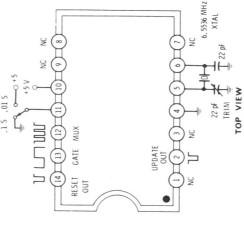

TIME BASE FOR FREQUENCY COUNTER

This package provides the time base and "housekeeping" for a frequency counter.

Pins 5 and 6 form a crystal oscillator as shown. The Gate output will be high for 0.1 second and low for 0.1 second if pin 11 is grounded, and will be high for 0.01 second and low for 0.01 second if pin 11 is high.

The MUX output is a 1.6-kHz square wave useful for multiplexing displays. The Update output is a brief negative-going pulse coincident with the rising edge of the gate output. It is used to transfer a count to display latches. The Reset output is a brief negative-going pulse that follows the Update output. It is used to reset a counter.

Operating current is 0.25 milliampere from a 5-volt supply.

7200
(INTERSIL)

WRISTWATCH

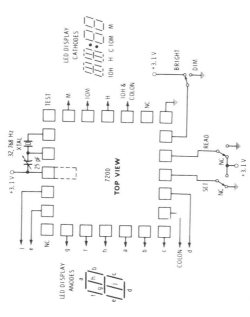

This package contains an entire wristwatch circuit, less only display, crystal, trimmer, control push buttons, and two 1.55-volt silver-oxide batteries. It is connected as shown.

Pressing Read once gives the time in hours and minutes. Pressing it again gives the day and date. Pressing it again gives a continuous readout of seconds for one minute.

Pressing Set once lets Read set the date. Pressing it again lets Read set the hour. The colon is off for a.m. and on for p.m. Pressing Set again lets Read set the day. Pressing Set again lets Read set the minutes. Pressing Set once more lets Read hold the seconds at 00 until one second after Read is released. Usually Read is released on a time tone.

A 10K resistor may temporarily be connected from Test to a 1.55-volt source (usually the center tap of the two cells). The 1-hertz square wave appearing across this resistor is used to adjust the trimmer capacitor to get the correct long-term accuracy. This is done with the display off.

Operating current with the display off is 4.0 microamperes. Display current varies from 20 milliamperes with 2 segments lit to 42 milliamperes with 7 segments lit. Oscillator stability is 1.3 parts per million, equal to 24 seconds per year.

Note that 9-segment readouts are needed for the hours positions if the day is to be read out.

Index

Optoelectronic devices, 10-13
OR gate, 9-29

Peak current, 4-5
Peak voltage, 4-5
Photoconductive cells, 10-13
Photodiodes, 10-19
Photometric system, 10-9
Phototransistors, 10-24
Photovoltaic cells, 10-15
Pinch-off voltage, 7-8
PIN diode, 4-17
PNP configuration, 5-5
Power temperature curve, 3-7
P-Type semiconductors, 1-21
Projected peak value, 4-7
PUT, 8-35

Radiance, 10-9
Radiant flux, 10-8
Radiant energy, 10-8
Radiant power, 10-8
Radiometric system, 10-8
Rectification, 2-31
Reverse-bias, 2-11

Schottky diodes, 4-18
SCR, 8-5
Schematic symbols, chart A-3
Semiconductors, 1-9
Seven-segment displays (7-segment), 10-38

Silicon controlled rectifiers, 8-5
Silicon, 1-9
Silicon diode, 2-17
Solid-state components, 1-5

Testing
    bipolar transistors, 5-29
    NPN transistors, 5-29
    PNP transistors, 5-30
Thin-film IC, 9-17
Thick-film IC, 9-18
Thyristor, 8-3
Transferred electron effect, 4-19
Tunner diode, 4-5

UJT 8-35

VI curve, 8-9
Valley current, 4-5
Valley voltage, 4-5
Varactor diode, 4-11

Zener current, 3-5
Zener diodes, 3-5
Zener knee impedance, 3-14
Zener test current, 3-7
Zener impedance, 3-14
Zener regulator circuit, 3-17
Zener voltage, 3-6
Zener voltage temperature coefficient, 3-11